MOTHS
a complete guide to biology and behavior

David C. Lees
& Alberto Zilli

Smithsonian Books
Washington, DC

ACKNOWLEDGEMENTS Special thanks go to our colleagues: Alessandro Giusti, Ian J. Kitching, Blanca Huertas, Geoff Martin, Klaus Sattler, Mark Sterling (the last two for extensive reviews and comments), David Wagner, Robert Hoare (official reviewers), Sandra Doyle (two figures with extensive corrections), Tim Littlewood, Erica McAlister (for discussions), and the editing and publishing teams. The following were especially helpful in sourcing suitable images (see also Picture Credits): Patrick Clement and John Horstman. The authors are also deeply thankful to the following for their kind help and assistance in various requests often at short notice: David Agassiz, Henry Barlow, Patrick Basquin, Joaquin Baixeras, Franziska Bauer, Gunnar Brehm, Gavin Broad, Peter Buchner, John Chainey, David Demergès, Anthony Galsworthy, Michael Geiser, George Gibbs, Pavel Gorbunov, Bob Heckford, Maria Heikkilä, John Horstman, Carlos Lopez-Vaamonde, Wolfram Mey, Joël Minet, Paolo Mazzei, Daniel Rubinoff, Suzanne Ryder, Mark Shaw, Giorgio Venturini, Eglė Viciuvienė, Roman Yakovlev, and Shen-Horn Yen. For their great support and understanding AZ acknowledges also Martin and Oliver Zilli, and not least Flavia Pinzari.

Published in Great Britain by the Natural History Museum,
Cromwell Rd, London SW7 5BD

Design by Mercer Design, London
Reproduction by Saxon Digital Services, Norfolk

Published in North America by Smithsonian Books

This book may be purchased for educational, business, or sales promotional use. For information, please write: Special Markets Department, Smithsonian Books, P.O. Box 37012, MRC 513, Washington, DC 20013

ISBN 978-1-58834-654-4

Library of Congress Cataloging-in-Publication Data

Names: Lees, David Conway, 1959- author.
Title: Moths : a complete guide to biology and behavior / David Lees and
  Alberto Zilli.
Description: Washington, DC : Smithsonian Books, [2019] | Includes index.
Identifiers: LCCN 2019020416 | ISBN 9781588346544 (paperback)
Subjects: LCSH: Moths.
Classification: LCC QL542 .L426 2019 | DDC 595.78--dc23 LC record available at
https://lccn.loc.gov/2019020416

Printed in China by Toppan Leefung Printing Ltd, not at government expense

23 22 21 20 19   5 4 3 2 1

Contents

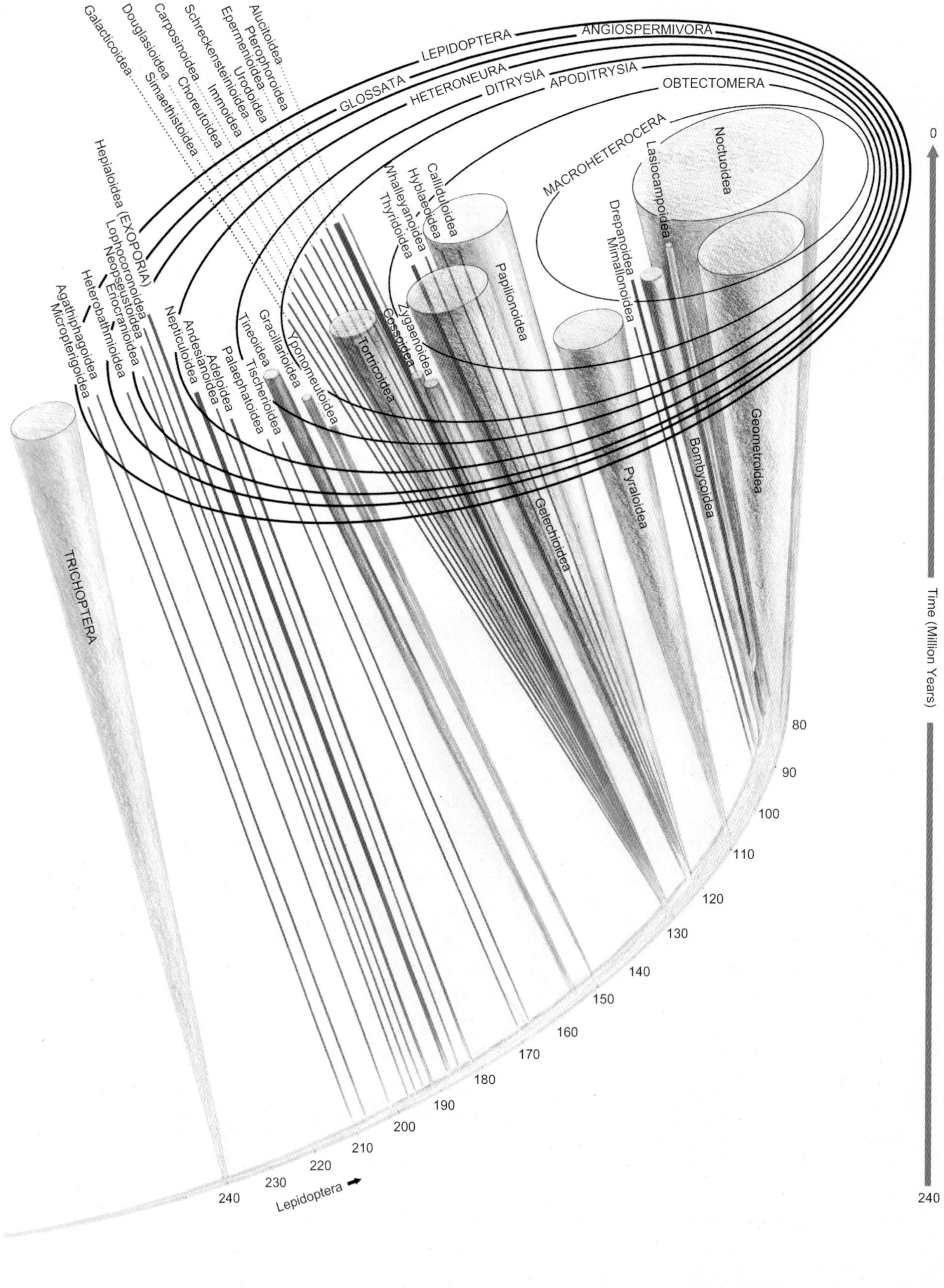

LEPIDOPTERA
ANGIOSPERMIVORA
GLOSSATA
HETERONEURA
DITRYSIA
APODITRYSIA
OBTECTOMERA
MACROHETEROCERA
Alucitoidea
Pterophoroidea
Epermenioidea
Urodoidea
Schreckensteinioidea
Immoidea
Carposinoidea
Choreutoidea
Douglasioidea
Simaethistoidea
Galacticoidea
Hepialoidea (EXOPORIA)
Lophocoronoidea
Neopseustoidea
Eriocranioidea
Heterobathmioidea
Agathiphagoidea
Micropterigoidea
Nepticuloidea
Andesianoidea
Adeloidea
Palaephatoidea
Tischerioidea
Tineoidea
Gracillarioidea
Yponomeutoidea
Tortricoidea
Cossoidea
Zygaenoidea
Thyridoidea
Whalleyanoidea
Hyblaeoidea
Calliduloidea
Papilionoidea
Gelechioidea
Pyraloidea
Mimallonoidea
Drepanoidea
Lasiocampoidea
Bombycoidea
Noctuoidea
Geometroidea
TRICHOPTERA
Time (Million Years)
0
240
80
90
100
110
120
130
140
150
160
170
180
190
200
210
220
230
240
Lepidoptera

INTRODUCTION

What is a moth?

I N COMMON CONCEPTION, MOTHS ARE FURRY, velvety insects that are mysterious denizens of the night. Moths have a potentially fatal penchant for artificial lamps or scurry about with expensive tastes in clothes made of natural fabrics or for tapestries and carpets. Historical and popular culture embodies moths as besotted with candles and allured by the flame, seemingly unbothered about scorching their wings. Moths appear to have been bewildered ever since Edison invented the artificial lamp. They are even something sinister, fluttering bat-like from dark places like spirits from the underworld. Butterflies, by contrast, come from the bright side of life.

Here, we attempt to untarnish their reputation by putting moths, with their astonishing diversity, firmly back in the limelight. Many moths also fly by day, and some outshine the glamour of any butterfly, while some butterflies, notably the 'browns', are drably coloured. As diurnal organisms ourselves, we tend to emphasie what we notice by daytime.

So what is a moth? A simple question but without a simple answer. It is best to begin by comparing moths with butterflies. English speakers, in particular, semantically distinguish these two very popular insect groups, which together make up the Order Lepidoptera (a name literally meaning scale-winged insects). Like several other insect orders, Lepidoptera go through four principal life stages. It is not in the egg, caterpillar or in the pupal stage that Lepidoptera have their keynote characteristic, however, but in the adult stage, where thousands of microscopic scales overlapping like tiles cover their wings. In adult Lepidoptera, there is unfortunately no one feature that can reliably serve to separate butterflies from moths. Two striking features, which are rules of thumb only, are clubs at the antennal tips and, better, the lack of a linking bristle at the base of the hindwing.

Answering what a moth is first begs the question, what is a butterfly? Before 1986, the common concept of butterflies turns out to have been wrong, as it actually did not include all butterflies. Many people even mistakenly separated the main families of butterflies from the skippers (family Hesperiidae). Many text books thus had 'butterflies and skippers' in the title. Others condescended that the skippers were 'mere' moths. In 1986, an obscure group of mainly night-

OPPOSITE Reconstruction of the evolutionary diversification of the Lepidoptera into its current 42 superfamilies (ending '-oidea'), depicted like a suckering underground branching system. At 'ground' level, extant moth diversity (the present day time slice) is indicated. The vertical axis represents time in millions of years ago. Diameters of today's cross-section reflect the number of described species in each superfamily of Lepidoptera compared with Trichoptera (caddisflies: the green cone). The major subdivisions of Lepidoptera (in capitals) are circumscribed by progressively inclusive black oval lines. The superfamilies are currently estimated to have successively split off the main root of the Lepidoptera at approximately the times indicated. The butterflies (Papilionoidea) (greyed out) stand as just a single branch within Ditrysia, which includes almost 99% of species of Lepidoptera; any other Lepidoptera group are moths (orange cones). Moths have thus astonishing diversity relative to butterflies. Note that the basal branching pattern of Apoditrysia and Obtectomera still remains to be resolved. (Extinct superfamilies are not shown.)

ABOVE Which is the butterfly and which is the moth? A trick question! The specimens are opposite sexes of the same species of a moth from Madagascar. The male (left) of *Pemphigostola synemonistis* (Noctuidae: Agaristinae) has clubbed antennae, just like a butterfly. The more colourful female (right), has thread-like antennae, and looks much more like a day-flying moth. Notice the warped forewings of the male, which has transparent structures at the base, used in sound production.

flying 'moths' from Central and South America, the family Hedylidae (with its only genus *Macrosoma*), was shown to be a group of butterflies. Their early stages are butterfly-like: the egg is tall and spindle shaped, like that of an Orange Tip; the caterpillar has a cleft tail recalling a Purple Emperor; the pupa is attached by a silken girdle, as in Pieridae (cabbage white family); and the adults have some subtle internal characters suggesting butterfly affinities. Adult hedylids, however, externally resemble loopers (Geometridae), with thread-like or even feathery antennae; the tip is not club-like as in many butterflies, nor crochet-hooked like in many skippers (Hesperiidae). Recent studies, particularly those based on DNA, have shown that butterflies, comprising hedylids, hesperiids and five other families, finally form a 'natural' group (a real entity in evolution), collectively known as the superfamily Papilionoidea. DNA studies also show that skippers and hedylids are each other's closest relatives.

In addition to the butterflies (Papilionoidea), according to current classification, there are another 41 superfamilies (all moths) constituting the Order Lepidoptera. Moths are non-papilionoid Lepidoptera, and they are astonishingly diverse. They make up 129 families of Lepidoptera and at least 90% of the known species of Lepidoptera, probably more. This dazzling diversity of moths can be seen in fundamental differences in body structure.

The vast majority of Lepidoptera, including butterflies, show a particular feature of the female sex, bearing two genital openings, one for mating the other for egg-laying. Species belonging to this group of some 30 superfamilies (with maybe 99% of moth species) are known as the Ditrysia. Moths also include five primitive superfamilies, the Monotrysia, that have only one opening for both functions, and a few other primitive groups have other female genital combinations too. There are thus many moth superfamilies more closely related to the butterflies than to primitive moth groups. Nonetheless, we call all non-papilionoid Lepidoptera 'moths', a descriptive term of convenience (as for 'fish' for example) not reflecting a natural group in the evolution of Lepidoptera.

WHAT ARE THE CLOSEST RELATIVES OF LEPIDOPTERA?

Despite their strong dissimilarity, either as adults or larvae, moths and butterflies (order Lepidoptera) appear to be most closely related to the caddisflies (order Trichoptera), to such an extent that zoologists often group them into a 'superorder' Amphiesmenoptera. Amongst their most important features in common, they have wings clothed by minute cuticular processes (termed hairs in Trichoptera and scales in Lepidoptera). Also, female Amphiesmenoptera have different sex chromosomes (XY), whereas in most other insects and animals, females are XX and it is the male that has different sex chromosomes. Interestingly, a third group of extinct Amphiesmenoptera, the Tarachoptera, was recognized in 2017 from Burmese amber dated to the mid-Cretaceous, around 100 million years ago. It is generally assumed that Amphiesmenoptera evolved during the Middle to Late Jurassic period around 225–240 million years ago, while the origin and early diversification of Lepidoptera would have occurred in the early Jurassic, around 180–200 million years ago, but these dates are subject to new fossil discoveries.

The current view of the evolutionary branching sequence of Lepidoptera is shown on p. 4. In a zoological classification, every superfamily has the same importance (i.e. they have the same rank). As illustrated though some superfamilies have deeper and other shallower connections beneath the surface that represents the diversity we see today.

Why write a book on moths? Well, there are many books on butterflies and those often gaudy insects usually steal the limelight. Moths make up what is missing, and there also many books on them, but here we paint a broad new portrait that integrates modern research findings, many of which are largely unsung. This book starts with an explanation of the main jargon and concepts in the world of moths, explaining how they are structured. Then, in seven more chapters, we explain how the adult moths come about, their often weird feeding habits and how they seek a mate. We unravel the extraordinary armoury of moth warfare, explore their biodiversity, ecology and distribution, and how they evolve. Finally, we shine a spotlight on their fascinating, even essential interactions with humans. Throughout, we tell you some of their most extraordinary life stories. Many of these examples demonstrate that moths are not only popular and fascinating, but serve among the foremost model systems of any insect that illuminate important concepts in biology.

LEFT Species of the genus *Macrosoma* from Central and South America (illustrated is *Macrosoma heliconiaria*) were long thought to be moths, and not surprisingly were grouped among the loopers (Geometridae) because of their striking resemblance with members of this family, but they are actually butterflies (family Hedylidae).

CHAPTER 1

Blueprint for success

JUDGING BY THE EXTRAORDINARY DIVERSITY of forms that have evolved and by which they have populated the Earth, the Insects (Insecta) are the most successful arthropod group. Entomologists have described over one million species of insects so far. The most conservative estimates set the total number of currently living insects at five or six million species (other projections indicate up to 30 million!). There are 29 extant main groups (orders) of insects, but the bulk of this diversity is taken by only four of them, namely Coleoptera (beetles), Diptera (flies, mosquitos and midges), Hymenoptera (sawflies, bees, wasps and ants) and Lepidoptera (moths and butterflies). Currently, beetles have the most known species but it is possible that wasps and flies have more species overall because they have not been sufficiently studied. Reasons for the evolutionary success of these four groups are varied, resulting from the overall success in colonizing terrestrial environments of a jointed exoskeleton to the acquisition of flight, in combination with complete metamorphosis. Here we focus on some reasons for the success of the Lepidoptera, from the perspective of both early and adult stages. However, we start with the basic blueprint for an arthropod, comprising features which are crucial steps in their success.

Arthropods (Arthropoda) are animals in which an external skeleton (or exoskeleton) combines with articulated legs and other appendages. The exoskeleton takes the form of a tight wetsuit, the cuticle, made of a waterproof elastic substance called chitin, which is hardened here and there by stiff plates of other substances, such as sclerotin in insects and calcium carbonate in crustaceans. Sclerotized plates (sclerites) may be locked together to enhance robustness or be separated by strips of chitinous membrane that thus enable mobility of body parts. The main functions of an exoskeleton are to protect the body and to allow locomotion via muscles inserted internally onto the body wall. It may seem obvious, but there is an important advantage of such a protective suit, in that it prevents water loss and insulates from the extremes of the environment. This is the main trick that allowed the insects to conquer, diversify and spread over terrestrial habitats. Being armoured in this way, however, meant that the exchange of respiratory gases on land had to follow another route compared to soft-bodied

OPPOSITE A freshly emerged male of Comet Moth (*Argema mittrei*; Saturniidae) in the rainforest of Manongarivo, Madagascar. The extraordinary long tails are bat deflection devices.

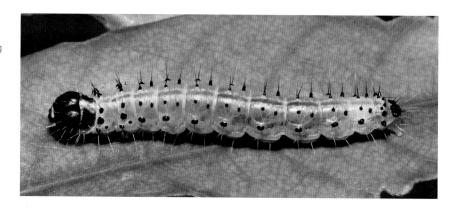

and aquatic organisms. In insects, oxygen intake and carbon dioxide release occurs via a complex system of tubes (tracheae) that branch off into the body from a series of valves (spiracles) on the cuticle.

Insects, as with all arthropods, are segmented animals. Segmentation is an effective way of making an organism via a simple building plan, i.e. the repetition of a single unit. Just line up several identical segments, seal with cap-like segments at the top and the bottom, and the basic bodywork is ready. It is thought that ancestral arthropods bore a pair of articulated appendages for every segment of the trunk. You can imagine them like modern centipedes. During arthropod evolution and diversification, however, some segments started to amalgamate and became specialized to fulfil particular functions. In insects, this process led their bodies to be organized into three different sections, the head, thorax and abdomen. The head segments, in particular, became strongly fused. Insects are therefore organisms in which the advantage of segmentation as a means of assembling an organism is coupled with that of specialization of various body parts so as to better accomplish an array of different functions. Furthermore, some segments lost their original pair of appendages, so that pairs only remained in the head, thorax and in some segments of the abdomen.

Strictly speaking, insects are arthropods with only three pairs of articulated legs, confined to the thoracic segments, and which have external mouthparts for the manipulation of food, unlike springtails, proturans and two-pronged bristletails, which have these parts inside the mouth.

The most outstanding novelty allowing an early evolutionary radiation by insects was the development of wings on the thorax. Wings represented an incalculable boost towards dispersal and colonization by insects over all terrestrial habitats. Though most insect groups are winged, they did not all attain the richness in terms of the number of species (in the hundreds of thousands) seen in the four giant insect orders mentioned above. These hyperdiverse groups share another key adaptation: they all undergo a complete metamorphosis. They comprise the vast majority of so-called Holometabola, which completely reshape their body during growth unlike other insects, such as bugs and grasshoppers, that have adult-like 'hoppers'. Following the hatching of the egg, holometabolous insects initially spend a part of their life as wingless larvae. Then they enter a pupal phase, during which their larval body is eventually converted into the adult configuration. The advantages of such a complicated process include the fact that Holometabola have different life stages that can be streamlined for different functions, for example feeding, reproduction and dispersal, to more successfully exploit their environments. When their essential function is to grow and develop, they are larvae, i.e. basically 'eating-machines' whose body is best suited for that role. When their essential functions are to reproduce and disperse, they are adults. When it is useful also to have a stationary stage in which developmental transformations can take place, they are eggs or pupae (in butterflies, the latter often known as chrysalids). They no longer need to compromise with a similar body plan during all their life cycle.

As regards the Lepidoptera, another two main factors have greatly promoted their diversification. The first is strongly connected to the wings, in terms of how these are composed. In fact, as their name indicates (from the Greek *lepidos* = scale and *pteron* = wing), the Lepidoptera are insects whose wings are covered by scales (scales are rare in other insect orders). These scales are of microscopic size (0.04–0.8 mm or 0.0015–0.003 in) and overlapping like tiles. On other parts of the body, scales are modified into hairs, or form tufts and brushes, imparting that familiar fluffy charm that moths are renowned for. The mosaic of scales is responsible for the astonishing variety of colours and patterns shown by butterflies and moths. Such colour patterns represent another feature that the forces of evolution can act on.

The second diversification factor relates to the particular feeding associations shown by the Lepidoptera. In fact, the overwhelming majority of species feed on plants (they are phytophagous). Whether via leaves or other parts, the plant kingdom is the main food source for caterpillars, as the larvae of Lepidoptera are widely known. Furthermore, most adult Lepidoptera rely on plants for nutrition. They suck nectar, dew or sap and (more rarely) digest pollen. These insects have then a double bond to the plant world, and what we know about the evolutionary diversification of the Lepidoptera is that it closely couples with that of flowering plants since the Middle Cretaceous period, from around 125 million years ago (see Chp. 6). Without the radiation of flowering plants, we would not have the current breathtaking diversity of Lepidoptera. And vice-versa, the diversity of flowering plants is due in no small part to the important role of Lepidoptera as pollinators. The range of defensive chemicals in leaves, so important to us for taste and medicine, is very much a response to the ravages of caterpillars, fossil traces of which date back well over 100 million years. In a broad sense, Lepidoptera and plants have coevolved.

BELOW Large Elephant Hawk (*Deilephila elpenor*; Sphingidae). The flexion of its wings in flight imparts great manoevrability.

THE MOTH'S HEAD

A moth's body is typically organized into three different sections: the head, thorax and abdomen. The head is the part where the primary subdivision into segments is least visible (see p. 10). Nonetheless, indications deriving from the number of appendages present and embryological studies suggest the original number of head segments was six or seven. The upper (hereafter dorsal) part of the head is called the vertex, and may be conspicuously tufted. Likewise, the forward part (the anterior) is the frons, which may be tufted or bare; some species sport conspicuous frontal protuberances, more strongly expressed in species that pupate in the ground, particularly in hardened soils, where adults emerging from subterranean pupae, such as the Pine Processionary Moth, need to force their way out. Despite the head not being particularly big relative to the thorax and abdomen, it is packed with appendages and sensory organs used to sense their environment and to feed.

RIGHT Close up of a gelechioid micromoth head showing the large compound eye with its many facets, the rolled up proboscis which is strongly scaled towards its base, the slightly upcurved labial palps and, on top, the antennal bases.

FAR RIGHT The head of *Janthinea friwaldszkii* (Noctuidae), showing the hardened spade-like projection, the dark plate between the eyes, which presumably helps the emerging moth break through hardened ground.

ANTENNAE

The antennae are possibly the most striking feature of a moth's head. If they do not bear distinctive clubs towards their tips, there is already a high chance you have a moth (read on though for exceptions). The paired antennae consist of two basal articles (the scape and pedicel), followed by up to several tens of jointed segments that altogether constitute the flagellum. In many groups, the scape bears a comb of bristles (pecten), or may be conspicuously expanded into an eye-cap, thus called because it at least partially covers the compound eye. This enables some minute moths, such as Nepticulidae and Lyonetiidae, to disguise their head outline. Inside the pedicel is the Johnston's organ, a bundle of sensory cells that detect the position taken by the flagellum, especially in flight.

The flagellum varies in shape. It is often assumed that butterflies are the only Lepidoptera with thin, apically clubbed antennae (hence Rhopalocera or clubhorns), whereas moths show other types of configuration (Heterocera). In reality, there are many moths (such as burnet and sun moths) whose antennae are essentially clubbed. A remarkable case is that of an agaristine moth, *Pemphigostola synemonistis* from Madagascar (p. 6), in which the male antennae are clubbed and the female ones thread-like. Moth antennae may thus be filiform (thread-like), rod-like, beaded, thickened towards middle, apically pointed, hooked, clubbed,

TOP LEFT A yet undescribed sun moth species (*Synemon* sp.; Castniidae) from Western Australia showing its strongly clubbed, 'butterfly-like' antennae.

TOP RIGHT Ornately adorned antennae sported by a female Red-banded Fairy Longhorn Moth (*Nemophora rubrofascia*; Adelidae) from subtropical China.

BOTTOM LEFT A male Gypsy Moth (*Lymantria dispar*; Erebidae) showing the fine pectinations that enhance the surface area available for detecting sparse molecules of the female lure (pheromone).

BOTTOM RIGHT A male *Amphoraceras jordani* (Erebidae) from New Guinea showing its complex antennae, its basal segments greatly narrowed and widened to wrap around the longitudinal axis, leaving an opening in the upper part in the form of a pitcher plant, presumably enhancing reception of female pheromones.

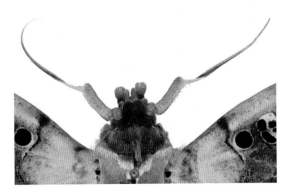

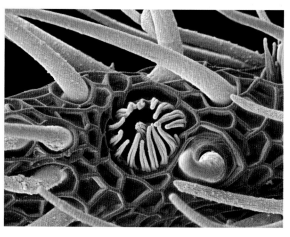

ABOVE Pupa of the longhorn *Nemophora degeerella* (Adelidae). In order to accommodate the extraordinarily long male antennae, the antennal sheaths spiral eight times round the abdomen. Emergence of the adult must be a delicate process to avoid breakage.

ABOVE RIGHT False-coloured image based on a scanning electron microscope (SEM) photo of a type of olfactory sensilla (sensillum coeloconicum) on a moth's antenna, measuring only about 10 μm across, set within a fluted surface. A conical peg cone is surrounded by a number of pickets.

or with side branches. In the last case, the individual segments of the flagellum bear one to four projections, so that the antennae are toothed (serrate) or comb-like (pectinate), giving them a feathery appearance. Sometimes two different configurations alternate along the antenna, e.g. comb-like to the middle, then filiform, like the antennae of the Leopard Moth (*Zeuzera pyrina*). The dorsal side of the antennae is normally the one clothed with scales, while the underside (ventral side) bears setae and other sensory organs.

The antennae are the 'nose' of the Lepidoptera and with them, moths smell (or taste) volatile chemical substances, sometimes in incredibly low concentrations. When the antennae vary in structure between sexes (i.e. they show sexual dimorphism), it is the males which generally have the most elaborate ones, because the males must detect the female calling lure from a distance. This is an important difference from butterflies whose females are not known to use chemicals as lures. Many of the looper family Geometridae (subfamily Ennominae, like the Peppered Moth), for example, have males with highly feathered antennae, which in the females are thread-like. The longest antennae in relation to body size belong to some male Adelidae. The males of the 2.5 cm (1 in) wingspan *Nemophora polychorda* have antennae up to 4.5 cm (1¾ in) long. They may be up to seven times the length of the body and it is really hard to imagine how the emerging moth can extract its antennae without breaking them from an antennal sheath when it is coiled many times around the tip of its pupa.

Most bizarre are also the antennae of *Amphoraceras jordani* (p. 13), a large moth from New Guinea. In this species, the 'amphora' is a vase-like structure only existing in males constructed like a wrap from the widened segments at the base of the flagellum, doubtless to detect the female perfume. In the same family (Erebidae), particularly noteworthy are moths in the subfamily Herminiinae (e.g. species of *Renia*) in which, always in males, some antennal segments are variously flattened, elongated and displaced from the longitudinal axis so as to produce a sort of clamp. Unfortunately, the courtship of these nocturnal species has never been observed, but it seems likely that the clamps serve to clasp the females.

CHAETOSEMATA

There is another sensory organ that is often present on a moth's head, called the chaetosema (usually one per side, or rarely a big fused one). Also known as Jordan's organs after the name of their discoverer (the eminent lepidopterist Karl Jordan), chaetosemata appear as raised areas, thickly covered in sensory bristles. Some primitive lepidoptera may show additional chaetosemata. They are easy to observe in some Zygaenidae. Their function remains insufficiently known.

EYES

Behind each antenna, and almost touching the moth's large compound eye, many species bear a small lens called the ocellus. The job of the ocellus is primarily to detect changes in light intensity. Together with the antennae the large compound eyes are usually the most conspicuous parts of a moth's head. The term 'compound' pertains to their structure, which typically consists of many near conical subunits, the ommatidia. Almost perfectly hexagonal when seen from above, the ommatidia are usually very numerous and in some hawkmoths they number as many as 30,000. The internal anatomy of the eye is complex and its exact configuration affects how the image is formed. Broadly speaking, many day-flying Lepidoptera, notably most butterflies with the exception of hesperiids and hedylids, and a few essentially diurnal families of primitive day-flying moths, have 'apposition' eyes. In this type, every ommatidium is shaded, from the surrounding ones, by dark pigment so that its internal sensory part (the rhabdom) can only be triggered by rays entering straight.

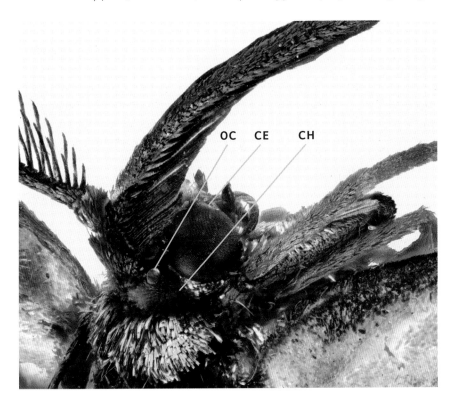

LEFT A closeup of the procridine moth *Clelea syfanica* (Zygaenidae), with red arrows pointing to chaetosema (CH), compound eye (CE) and ocellus (OC).

RIGHT Diagram of apposition eye (left) and superposition eye (right).

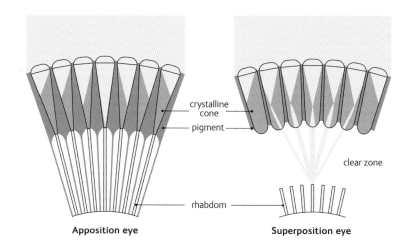

crystalline cone

pigment

clear zone

rhabdom

Light

Apposition eye

Superposition eye

The overall image in apposition eyes is built by all the individual optical elements, not unlike a mosaic. This kind of eye results in sharp vision that is especially sensitive to movement, but to be effective it needs plenty of light. Curiously enough, it produces a diffuse polka dot effect of darker spots visible to us on the compound eye.

Species flying at night and even many day-flying species that also fly at night cannot always rely on strong light intensity. These have 'superposition' eyes that are adapted so as not to waste any of the few precious rays available nocturnally - the system used in the vast majority of Lepidoptera. Their ommatidia are not shaded for the entire length from adjoining ones, so that there remains a clear zone where each ommatidium can collect light from neighbouring ones. The light entering at different angles from the ommatidia surrounding a central one will thus also hit the rhabdom in question rather than being absorbed by their walls. Furthermore, the more external part of each ommatidium consists of an increasingly dense medium that has a lensing

RIGHT Glowing eyes of Pandora Sphinx (*Eumorpha pandorus*; Sphingidae).

effect on the rays, which are bent and converge on the sensory parts from adjoining ommatidia. In darkness, the superposition type of vision has been estimated to be up to 1,000 times more effective than apposition, at the price of losing much of its resolution (just as higher ISO in photography allows imaging at low light but reduces sharpness). As research advances, however, several intermediate systems and mixed combinations are being discovered in various moths' eyes depending on whether they are diurnal or nocturnal, or both. Hawkmoths can recognize and learn colours even in low light (like starlight) using receptors tuned to UV, blue and green. When a moth is illuminated by torchlight the eyes tend to glow, often yellow to red. This is due to the reflection of light by air trapped in a dense reflective tracheal network forming a layer in the inner part of the eye known as the tapetum. Visibility of the tapetum can be enhanced once a moth has been accustomed to darkness for a while. In fact, it has been shown that in superposition eyes the pigment granules of the walls shielding light from nearby ommatidia can migrate from deeper to shallower parts of the ommatidium, thus increasing the glowing effect by exposing better the clear zone.

The overall shape of the moth eye is generally globular, but there are some unusual configurations, such as the elliptical eyes of several day-flying species (e.g. species of *Schinia*), and the greatly enlarged eyes of some swarming males that need to spot females quickly (e.g. *Nemophora* longhorns). Perhaps most surprising to find in Lepidoptera are eyes split into two, a superior and an inferior one, a feature more familiar for diving beetles but occurring also in males of some Roeslerstammiidae (thus called Double-eye moths). An extreme case is a *Telethera* species from Sulawesi, where the two 'half-eyes' are shielded from one another by a fan of scales. In the New Guinean uraniid moth *Alcides agathyrsus*, a difference in eye size is noticeable between males and females, the latter having relatively small heads and eyes. Surprisingly, there is huge variation in eye size between males of this species. Some have nearly 'holoptic' eyes, that is, they are large enough to meet near the midline, as commonly seen in hoverflies. The lepidopterist Josiah Obadiah Westwood from the Hope Museum, Oxford even named one example of this species as '*boops*', literally meaning 'cow-eyes'. Many moths have hairy eyes, that is, they have hairs or setae between the ommatidia, such as found in many hadenine noctuids, where they might have a sensory function. A few moths have prominent eyelashes partially covering the eye. An example is both sexes of shark moths of the genus *Cucullia*, or in males only of *Dahlia* species (Erebidae: Herminiinae) from Australasia, in the last case originating from the antennal base.

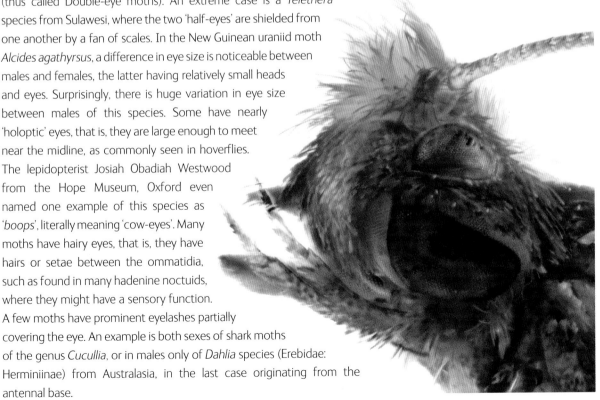

BELOW Double-eye moth from Sulawesi (*Telethera* sp.; Roeslerstammiidae), showing the bizarrely split compound eye of the male. The condition may be related to aerial behaviour when these moths swarm.

WHY ARE MOTHS ATTRACTED TO LIGHT?

The potentially fatal lure of the moth to the flame has long been a scientific mystery. Most nocturnally active moths are attracted to light, a phenomenon known as positive phototaxis. However, some species like the Old Lady (*Mormo maura*) tend to be repelled by it (they are negatively phototactic). With the invention of ultraviolet (UV) lamps for medical purposes just before the WWII, it was discovered that sources rich in UV greatly increased moth attraction to light. Insects, and especially moths, are particularly sensitive to the UV part of the electromagnetic spectrum. There have been a number of theories that try to explain this.

A common theory is that moths are attracted to the moon, and therefore they should fly higher on moonlit nights. A better theory is that moths can use the moon or stars to orientate, and that a moth adjusts its flying track to keep the light source at a constant angle to the eye. However, whilst rays from a celestial source would all be seen as parallel, those from a lamp radiate all around.

Accordingly, a moth on the wing would constantly turn inwards to keep itself at a constant angle to the light, ending up in a spiralling path which would make it eventually collide with the lamp. However, moths rarely exhibit such geometric trajectories, but rather take circuitous routes when coming to light, making loops and coils perhaps due to a compromise with escape responses or disturbance by wind plumes. Moths are also affected by a general phenomenon known as dorsal light reaction. Most flying animals, in fact, tend to keep the lighter sky above them (they do not fly upside down!), and will therefore also dip down when closing in on an artificial source that they then confuse with the sky light. Moth traps are designed to exploit the inwards spiralling responses of moths, using suitably placed barriers (baffles) around the lamp that they can collide with, so that they will then fall down through a collecting funnel into the trap.

In the 1970s, Philip Callaghan developed the infrared theory of light attraction. His view was that UV light pumped moth female pheromone molecules in the air into an excited state, so they emitted photons of infrared microwave radiation that could potentially be detected by sensilla on the male antennae, that he postulated were the right size to function as waveguides. The theory has not, however, gained much traction, because although males are more frequently attracted to light, it is known that pores on the moth sensilla are just the right size to detect pheromone molecules directly. Males are the more mobile sex anyway, and this hypothesis does not satisfactorily explain the attraction of females to light.

It is still not fully known how far you can attract moths from using an artificial light source. A classic experiment in 1978 by Robin Baker and colleagues at Manchester University suggested that most moths are attracted to light traps on the ground when they are only in the range of the light by just a few metres. Other trials in Germany in a region away from light pollution though, have shown that street lamps can attract moths only up to about 10–25 m (30–80 ft) away. In the last two cases, the light

LEFT Examining moths attracted to a sheet in Nouragues, French Guiana. The upper bulb is used to lure down higher flying moths from the rainforest canopy.

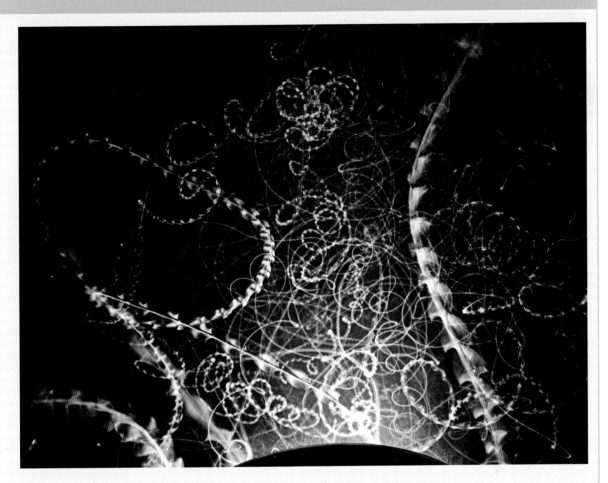

is only attracting the positively phototactic moths that happen to wander into the light source's small sphere of influence during the night.

Light attraction works better on dark nights or in places where there is no competition with other light sources, including notably the moon, and so a longer distance response should also work. The moth's behaviour of keeping a constant angle between its flight trajectory and the light rays emitted by an artificial source would then enable a moth to arrive at a single brightest light source perceived from a matter of kilometres as if it were a star.

Tropical biologist Daniel H. Janzen in a 1984 classic paper on big moths in Costa Rica noticed that many hawkmoths (Sphingidae) were feeding on flowers nearby a light source and yet not being attracted to it. Janzen asserted that sphingids behave differently through their adult life and spend some time building up a model of

ABOVE Moths make a series of crazy coils and spirals at a light, as captured by a multiple flash and slow exposure.

their home range early after emergence. At this time they rely on celestial cues and are more prone to be attracted at light. Once familiarized with their habitat, he proposed they could switch off their initial positively phototactic response and turn to orienting using landscape features. This switching mechanism needs detailed testing, but Janzen asserted that mainly fresh hawkmoths were found at light whereas those feeding on nearby flowers were often worn, hence older.

Not all moths will be attracted to light, and the reasons for positive and negative phototaxis are not clear. The latter has, however, a clear meaning for cave-sheltering species like the Tissue Moth (*Triphosa dubitata*), which actively flies towards deep dark patches among rocks.

MOUTHPARTS

Moth mouths are quite complicated. They reflect part of their evolutionary history in their feeding apparatus. Primitive moths, whose direct descendants are still alive today, although with a restricted number of species, appeared during the early Jurassic period over 180 million years ago. They had a biting-chewing apparatus like that currently seen in modern-day members of the families Micropterigidae, Agathiphagidae and Heterobathmiidae (the mandibles are functional only in adults of the first and last families). In *Agathiphaga* they are used just to bite the moth's way out of the cocoon. It is thought that such archaic moths fed upon fern spores or the pollen of gymnosperms (cycads, gingkoes, conifers), since flowering plants (angiosperms) had not appeared yet at that time. However, some micropterigids subsequently shifted to gnawing angiosperm pollen following the radiation of flowering plants during the Cretaceous around 125 million years ago. The mandibles are accompanied by other appendages for the manipulation of food, including a pair of maxillae, each one ending with two short lobes facing the mouth and a slender, (primitively five-) segmented outer maxillary palp. Above and below these parts, two transverse structures are found, termed the labrum and labium, respectively. The former underwent great modification and reduction in most modern moths, even splitting into two lateral lobes. The latter terminates also with paired lateral, multi-segmented labial palps. While these are very small (or absent) in primitive moths, in some more modern moths they are dramatically conspicuous, as in the 'snouts' (Erebidae: Hypeninae).

During the subsequent evolution of the Lepidoptera, these insects had increasing opportunities for finding liquid resources. Flowers started to populate the world, and with them came nectar! This was the most important evolutionary innovation by flowering plants for luring and rewarding pollinating insects. In order to explain how evolution works, we should not imagine landscapes filled with flowers hoping to be pollinated by insects which developed licking or sucking mouthparts, but rather an integrated process during which both plants and insects reciprocally diversified. As regards moths, there would already have been forms partly capable of imbibing liquids such as sap from plant wounds or other plant exudates, and not least water with essential nutrients diluted in it. These early forms would have taken advantage of a changing vegetational scenario, with modern Lepidoptera eventually stemming from them.

What happened to the mouthparts of these new lineages of moths? Essentially the chewing parts – the mandibles and maxillae – atrophied or were lost (not always together!) and a sucking organ developed. This is the well-known straw-like structure, the proboscis of moths and butterflies. This is typically rolled up under the head when at rest. Often long – up to 28 cm (11 in) in the hawkmoth *Amphimoea walkeri* – it is derived from the extraordinary elongation of the paired galea, the outermost of the two maxillary lobes. The elongated galeae zip up along their entire length via tiny linking hairs or interdigitating edges, forming an internal canal for sucking. The labium positioned underneath underwent a substantial remodelling too, especially with substantial enlargement of the palps. Usually three-segmented and ornately clothed with scales, the labial palps are one of the more conspicuous features of moths' heads, while the maxillary palps are generally reduced or lost. The primary function of the labial palps is that of bearing sensilla and protecting the coiled proboscis. Sometimes they can obfuscate the outline of a moth and even mimic the petiole of a dead leaf to perfect their camouflage. The labrum, in turn, always remained comparatively small, though it underwent important modifications and adaptations. Most frequently it split from a single component into paired hairy (setose) lobes. Due to the extensive diversification of modern Lepidoptera, however, proboscides (the correct term for proboscises) are not all the same, and their morphology and development reflects their different feeding habits, including fasting. In fact, many groups (such as saturniids, psychids, some tussocks, and most winter-active geometrids) do not feed at all during the adult stage and solely rely on the reserves accumulated during the larval period. Accordingly, such moths either have reduced and non-functional proboscides or have lost them completely.

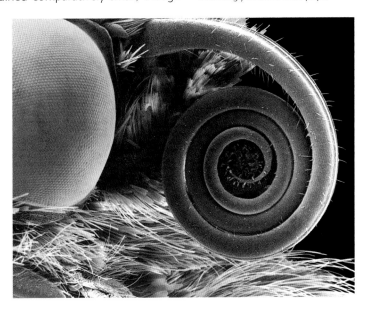

BELOW The rolled up proboscis of a noctuid moth (*Xestia dilucida*).

BOTTOM The handsome oecophorid moth (*Oecophora bractella*), showing its strikingly elbowed labial palps.

ABOVE Plain Golden Y (*Autographa jota*; Noctuidae), exhibiting its prominent thoracic tufts.

THE THORAX

The thorax consists of only three segments, called from front to back the prothorax, mesothorax and metathorax, yet it is voluminous. Each of these segments bears a pair of legs ventro-laterally. The last two also bear a dorso-lateral pair of wings (except in wingless forms, discussed below). Following a narrow membranous neck behind the head, we find the prothorax, the thinnest of the three segments as it does not need to accommodate powerful wing muscles. This segment can easily be spotted by a pair of densely scaled flaps called the patagia, often nicely decorated and evoking an Elizabethan ruff. Sometimes the patagia conceal a pair of glands underneath. The middle segment is the chunkiest part of the thorax. The reason for this is clear if we consider that it carries the first wing pair (forewings), which sustains most of the load during flight. The second wing pair (hindwings) is carried by the metathorax. Shielding the junction between the mesothorax and forewings are roughly triangular pieces like shoulder pads, the tegulae, whose colour pattern may be distinct or coordinated with the rest of the thorax. The dorsal part of the thorax, in turn, bears in some species complex scale tufts or crests. The thoracic trunk often bears organs involved in acoustic communication. Many moths possess a pair of 'ears', to detect the sound of bats, though they can also be used in communication. Those of one of the most speciose superfamilies, the Noctuoidea, are always found on the sides of the metathorax, although they can be hard to spot in fully scaled moths. Arctiine erebids also have (always on same segment) paired tymbals, specialized membranes that bear a narrow series of ridges that emit high frequency sounds when buckled by an underlying muscle, like the well known ones of cicadas used for sound emission.

WINGS AND SCALES

Wings, the most noticeable structures of the body plan of Lepidoptera, are used in manifold ways. They enable flight and its direct consequences (e.g. escape, dispersal, migration, foraging) but they are also used for many other functions, such as thermoregulation, camouflage, warning display and intraspecific communication such as courtship. Wing shapes, colours and patterns are engraved by the forces of evolution and often reveal the traces of contrasting demands. Just consider how with one and the same pattern some species must trade off the needs to be showy enough to attract partners (especially for diurnal species) and to be sufficiently camouflaged not to be noticed by predators. Often the upperwing and underwing are radically different, and usually the upperside is the most conspicuous. In collections, moths are rarely mounted upside down for this reason.

If we carefully remove the scale clothing from wings, they appear as transparent boards ridged with a number of stiff longitudinal and a few transverse 'veins'. The wings are broad, flat, cuticular protrusions of the body. The dorsal and ventral layers of epidermal cells of the wings are so closely appressed that no intervening space remains, except next to the framework of veins, where the nerves and tracheae are accommodated. The function of the veins in the fully expanded wings is essentially mechanical, to reinforce and to provide elasticity to the wing blade. However, during development, as the adult emerges from the pupa, the wings appear as short soft stumps, still to be inflated. At this soft stage, haemolymph (insect blood, which is usually transparent or yellowish) can still be pumped into the veins from the body cavity in order to expand the wings to their final size. Once this is achieved, the wings dry out (usually taking around half an hour, when they are generally held down parallel and vertical in response to gravity). Once hardened, the moth is ready for flight. If drying takes place before the wings are fully inflated, they will forever remain wrinkled and defective. That is why, soon after emergence, adults usually climb and hang from a support to enable the wings to spread unimpeded. This is a critical moment of the moth's life and one where it is imperative that there is no delay or obstacle.

The pattern of wing veins is an important feature for classification. For instance, primitive moths have an arrangement that does not differ too much between forewing and hindwing, and for this reason they are said to be homoneurous (from the Greek 'similar neuration'). On the other hand, butterflies and more recently evolved moths (i.e. after Exoporia, see p. 4) have a heteroneurous venation ('different neuration'), because veins of the forewing and hindwing strongly differ in number and branching details.

BELOW Magnified, ribbon-like arched scales of Sunset Moth (*Chrysiridia rhipheus*; Uraniidae). Its exquisite iridescent colours depend on the interaction between light and the fine ultrastructure of the scales.

BOTTOM A Black-veined Moth (*Siona lineata*; Geometridae), showing clearly its wing venation on the underside.

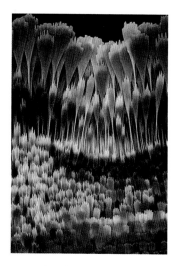

ABOVE Magnified portion of a wing of Silver Y (*Autographa gamma*; Noctuidae), showing plume-like scales along the wing edge.

As already seen, both the upper and lower surfaces of wings are extensively covered by scales. Strictly speaking, however, the moth's entire body is covered by scales. Scales consist of chitin and show diverse modifications in shape, from long, thin and hair-like – more usual for body scales – to broadly flat and spatulate. Flat scales prevail on the wings but are also commonly observed on the labial palps, vertex and thorax. In their most usual configuration, scales appear as flat blades that are inserted via a short neck (peduncle) into a socket onto the wing surface. But it is only under strong magnification that they show the intricate ultrastructure of the exposed surface, with numerous parallel ridges. If we push up the magnification, we can see micro-ribs alongside the ridges, bridges connecting the ridges, surface perforations leading to internal chambers and, depending on the particular moth species that we happen to examine, an endless variety of lamellae (micro-layers like plates in a car battery), crests and grooves sculpturing the surface. Along the outer and inner edges of the wings, furthermore, there are contouring scales that produce the so-called marginal fringes. Each scale is the product of a single specialized epidermal cell and may contain pigment that gives the scale a particular colour.

Pigment is not the only way in which scales bestow moths with colour. Scales also produce physical (or structural) colours, which derive from the play of light on the surface of the scales, including interference, diffraction and scattering. Scintillating iridescent colours are produced by interference. Chitin is arranged within scales in a

HOW DO MOTHS USE THEIR SCALES?

Lepidoptera are in effect clothed insects. The functions of scales go beyond that of providing moths with colourful patterns. Scales are fairly easily detachable, which is an essential feature when a moth gets trapped in a spider's web and needs to escape from its sticky threads. Some scales are fringed in order to disseminate scents that may be produced in different regions of the body. Black scales containing melanin help day-active species to absorb the sun's heat. More generally, a thick scale vesture insulates night-active moths flying in cold environments or seasons. Furthermore, scales are water-repellent and prevent moths from becoming soaked during rainfall. Scales absorb the sonar produced by bats, so that a descaled moth would be much more 'visible' to their most fearsome predators, bats. Bats spot their prey via echoes returning from sounds they emit, so a soundproofing coat helps absorb bat sounds and disrupt echoes, thus enhancing their chances of escaping predation. It is also worth noting that when we compare moths provided with hearing organs with those devoid of such organs, the earless moths of comparable size are often much fluffier. This fluffiness acts like an invisibility cloak for moths that cannot hear approaching bats.

ABOVE Some moths have highly elaborate thick scaling. A pyralid moth from French Guiana (*Pachypodistes* sp.) here shows its extraordinarily scaled legs like fancy boots.

series of tiny overlapping semi-transparent layers. Thus when light hits a scale, a wave is reflected but another wave goes deeper and enters the first layer before being also reflected, and so on for additional layers. Reflected rays are therefore slightly displaced in space and, depending on how they combine, they may constructively enhance a single colour, which then appears more vivid and shiny. A feature of these interference colours is that they depend on the angle at which light hits the surface. So if a moth changes the orientation of its wings, the colours may appear more brilliant or fade away.

BELOW The White Witch (*Thysania agrippina*; Erebidae), the widest moth in the world, is so large that it can hardly rest upright on a small trunk.

Sometimes, scale colours are not exactly what they seem to be. The late British lepidopterist W. Gerald Tremewan, who studied burnet moths, stressed the importance in distinguishing in genetical studies between 'true orange-coloured' forms and 'pseudo-orange' ones. The latter happen to look orange because of a mixture of yellow and red scales that are combined visually by us due to their small size, for example in the Narrow-bordered Five Spot Burnet (*Zygaena lonicerae*).

The size and shape of moth wings is incredibly varied. World record registers, books and websites are filled with speculation about which is the largest moth species in the world. In fact, wingspan – the distance between the tips of two forewings – should be measured with the specimen set in a standardized 'museum' position, with the hind margin of the forewings exactly perpendicular to the longitudinal axis of the body. If the forewings are reclined slightly further back, their tips would stand further apart. In the collections of the Natural History Museum, London there is a specimen of the White Witch (*Thysania agrippina*) from South America with a wingspan of 29.8 mm (12 in). Other gigantic species are members of the Indo-Australian atlas moth genera *Attacus* (particularly *A. crameri*, *A. caesar* and the Atlas Moth, *A. atlas*) and *Coscinocera* (in particular *C. omphale* and the Hercules Moth, *C. hercules*) which, despite unconfirmed records of single individuals reaching or surpassing that figure, have broader forewings and hindwings than *T. agrippina*, and therefore a larger wing surface area. Claims of moths with wingspans of up to 36 cm (14¼ in) though have not been confirmed and may actually derive from specimens not set in a fully standardized position.

RIGHT Species of the genus *Attacus* (Saturniidae) are among the record holders for wing surface area. Illustrated here is *Attacus taprobanis* from India, Kerala.

Regarding the smallest moth, similarly there have been disputes about the record holder. The smallest moth in Europe is probably the Sorrel Pigmy, *Enteucha acetosae* (Nepticulidae), which has a wingspan of 2.65–4.1 mm (0.10–0.16 in), according to a comment on ResearchGate by Thomas Sobczyk. There are examples of individuals from the same family collected in alcohol traps from California and the Congo reputed to go down to 2 mm (0.08 in) in wingspan, according to Lynn Kimsey. In fact, an unidentified specimen from the Congo at Bohart Museum collected by Steve Heydon has a forewing length measuring around 0.85 mm (0.033 in). However, the fringe appears to be missing on this specimen and thus an estimated full wingspan would be between 2–2.5 mm (0.08–0.1 in). The moth with the next smallest recorded and the smallest average wingspan is *Stigmella maya* from Belize. Its wingspan is 2.57–3.3 mm (0.1–0.13 in) with a forewing length of 1.4–1.5 mm (0.055–0.059 in) and this is currently the record holder considering average wingspan among described pygmy moth species. *Stigmella diniensis*, endemic to southern France, has possibly the smallest recorded forewing length of 1.3 mm (0.05 in), but other measurements go up to 1.6 mm (0.06 in), so further accurate measurements should be made for this species. *Trifurcula ridiculosa* from the Canary Islands was previously supposed to be a holder for the *Guinness Book of Records* at 2 mm (0.08 in). However, this figure clearly relates to its forewing length – measurements from museum specimens are 1.95–2.1 mm (0.077–0.082 in), but the wingspan ranges between 4.1–4.5 mm (0.16–0.18 in).

BELOW Sorrel Pigmy (*Enteucha acetosae*; Nepticulidae) one of the world's smallest moths.

Wing shapes can be bizarre. Perhaps the oddest are moths with deeply divided wings, which have wing blades tracing the wing veins as if they are spokes of a tatty umbrella. These include the many-plumed moth family Alucitidae recently combining also the false plume moths (Tineodidae), and the true plume moths (Pterophoridae). A popular example is the Twenty-plume Moth (*Alucita hexadactyla*), which actually has 24 individual 'plumes'. Other micromoths such as members of the families

LEFT Twenty-plume Moth (*Alucita hexadactyla*; Alucitidae) showing how deeply a moth wing can be divided.

Nepticulidae, Bucculatricidae, Gracillariidae and Cosmopterigidae have their hindwings greatly reduced to thin strips, but the overall surface area necessary to sustain flight is maintained, thanks to the exceedingly elongated hair-like scales of the fringes. Moth hindwings appear more prone to variation in shape than the forewings. This is because the latter play a greater role in thrust and lift, and are therefore subject to strong functional constraints that limit modifications. Some moths have normal forewings while the hindwings are reduced to small flaps. This is seen in the looper *Celonoptera mirificaria* and males of the erebids *Buzara chrysomela* and *Ophyx pratti*. Often female wings, especially hindwings, have a larger surface area (most effective in the downstroke) due to the need to bear eggs or even the weight of a male during mating, when the pair takes flight. Some moths, such as the Malagasy lithosiine moth *Phryganopteryx postexcisa* (p. 29), look as if a predator has taken a large bite out of each hindwing. A number of species bear elongated tails on the hindwings, notably swallow-tailed moths from the genera *Urania*, *Sematura*, and *Erateina*, as well as *Epicopeia hainesi*. Particularly remarkable in this respect are the tails of the saturniid moth genera *Argema*, *Actias*, *Copiopteryx*, *Eudaemonia* and *Antistathmoptera*, whose puzzling function has only recently been shown to be one of defence. The tails deflect bat attacks away from the moths' vital organs and towards expendable parts of the body, particularly the tails themselves, as their loss does not impair flight. Not only that, in *Argema* and *Actias*, the tips of the tails are twisted to deflect bat sonar in misleading directions. Male specimens of the Comet Moth (*Argema mittrei*), have the longest of all moth tails at around 13 cm (5 in). However, in relation to overall hindwing length, the 12 cm (4.7 in) long tails of *Eudaemonia trogophylla* and *E. argiphontes* from equatorial Africa are even more impressive. Members of the essentially day-flying family Himantopteridae (e.g. *Semioptila*) are also long-tailed and may even have the whole hindwing transformed into a tail. The wings then bear an extraordinary resemblance to those of the thread-winged antlions (Nemopteridae).

RIGHT Underside of the male (left) and female (right) of Clifden Nonpareil (*Catocala fraxini*; Erebidae) showing a characteristic common to many moths: the frenulum (bristles, arrowed, one in male and more in female), which links into a retaining retinaculum (hook/pocket) on the underside of the forewing base.

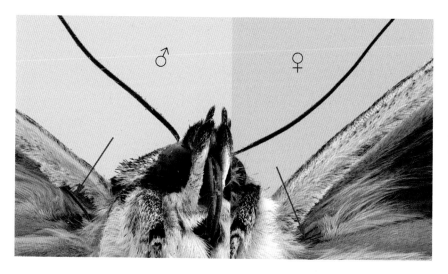

The process of beating is essentially carried out by the forewings, so the hindwings have to somehow coordinate with their pace, while expanding the moth's overall wing surface area. It is an advantage for both forewing and hindwing to behave as a single aerofoil. In most moths the hindwings are coupled with the forewings and remain slightly tucked under them to prevent a complete twist or overlap that would disrupt the airflow and impair flight. To achieve this, moths have developed a variety of special coupling mechanisms. There may be a row of short, hooked bristles that grapple one wing onto the other, or thickened veins on the two wings that overstep each other and prevent slippage, or a small forewing lobe (the jugum, seen in primitive moths such as Micropterigidae, but possibly of limited use e.g. in Hepialidae) linking fore- and hindwing, as well as several other systems. Most widespread, however, is the arrangement that involves a long bristle, the frenulum, which arises from the base of the hindwing and extends onto the underside of the forewing. As one of the more universal features uniting butterflies, note that the frenulum is absent in all butterflies except the night-flying hedylids and in the male of Regent Skipper (*Euschemon rafflesia*). Usually present in moths, the frenulum is retained by a hooking device known as retinaculum. Interestingly, with some exceptions, the sexes differ in the configuration of this wing-coupling mechanism. In males, the frenulum is usually a single stiff bristle that slips under the retinaculum, a small cuticular catch on the underside of the forewing towards its base. In females, the frenulum is multi-bristled and the retinaculum is a pouch composed of modified scales into which the frenulum tucks. Last but not least, another widespread method for ensuring wing-coupling is the amplexiform system, in which the two wings simply overlap to such an extent that the hindwing cannot slip off and overlap onto the forewing upperside (many Bombycoidea and most butterflies use this method).

It is noteworthy that some species seem to have developed sophisticated aerodynamic arrangements to increase flight performance. Hawkmoths are amongst the most powerful fliers and have forewings shaped rather like the swept-back wings

of fighter planes. Male moths especially, tend to have wings built for flight, and even in moths lacking pronounced sexual dimorphism, e.g. hawkmoths, wings of females may be significantly broader. Females have to carry more load due to their eggs but are generally less mobile, which means they have less chance to end up as supper.

WHY REDUCE YOUR WINGS?

Wings have facilitated moth diversification to such an extent that it seems to make no sense for moths to abandon such structures. Bizarrely, this is just what has happened in a number of species that show various degrees of wing regression that render them unable to fly. Wing reduction occurs mainly in the female sex, and there are just a couple of dozen examples where it has happened in both sexes. Some moths are stenopterous, in that they bear particularly narrow wings incapable of sustaining them during flight. This is the case for the female oecophorid, *Pleurota marginella*, or both sexes of the tortricid Sorensen's Agile Moth (*Sorensenata agilitata*), from New Zealand, the crambid *Exsilirarcha graminea* from New Zealand and the gelechiid *Catatinagma stenoptera* from Turkmenistan. These moths, with narrow forewings and reduced hindwings, are nevertheless good jumpers. They rapidly bounce among the vegetation or on the ground. More common are examples of the general phenomenon called brachyptery, where the wings are reduced overall, sometimes to tiny stumps, or full apterism, when individuals are completely wingless.

Reasons for the rare phenomenon of wing reduction, known in less than 1% of moths, are various. Broadly speaking, two main categories of flightless females are known, those that are in any case highly mobile, and those strongly sedentary. The first category, the spidery wingless females, for example the female of the Winter Moth (*Operophtera brumata*) can easily walk and crawl or climb, thanks to their agile

ABOVE Male of the Malagasy moth *Phryganopteryx rectangulata* (Erebidae). Some moth wings even though fully expanded exhibit irregular indentation or scalloping. Note that there are chunks apparently missing from the hindwings, which makes this probably poisonous moth look like a survivor of a previous predation attempt.

BELOW *Pringleophaga kerguelensis* (Tineidae) is a moth with highly reduced wings. There is little need for wings in a place as cold, bleak and windy as the Kerguelen Islands (also known as Desolation Islands), Antarctica.

legs, and some can make short airborne jumps if they still retain remnants of wings. The other category of flightless females, in contrast, tend not to disperse around and usually remain in the surroundings of their cocoon, like the Vapourer Moth (*Orgyia antiqua*), even if they have fully functional legs. In some extreme cases, however, females are also legless and cannot abandon their cocoons (female bagworms, some of which are even larviform). As many flightless forms have been recorded from isolated massifs, mountaintops, oases, islands or otherwise restricted habitats surrounded by unsuitable ones, it can be assumed that there is no advantage to leave the spot where a female's progeny may breed successfully. If any such female dispersed, in fact, she might easily end up in an environment hostile to her offspring. Under these circumstances, natural selection thus favours any adaptation capable of restricting the dispersal abilities of females. The fact that wing reduction is a highly unusual occurrence in males reflects their need to be fully mobile in order to look for mates. As males' reproductive success depends on encounters with female partners, they cannot give up their mobility and remain fully winged, even in windswept and barren environments.

RIGHT An example of partial wing reduction of Sorensen's Agile Moth (*Sorensenata agilitata*; Tortricidae) from Campbell Island, a subantarctic island south of New Zealand.

LEFT A mating pair of the fully winged male and the spidery, almost totally wingless female of the Winter Moth (*Operophtera brumata*).

Nonetheless, there are a few exceptional species, mostly from oceanic islands, in which males also are apterous, brachypterous or stenopterous, very probably because flight becomes too expensive to maintain fecundity in cold environments and to limit the chances of being wiped out by winds and thrown out to sea. Examples are again *Sorensenata agilitata* (Campbell Island), species of the tineid genus *Pringleophaga* including the Marion Flightless Moth (*P. marioni* from the South Indian Ocean), the enigmatic (possibly an yponomeutid) subantarctic *Embryonopsis halticella* (Kerguelen Islands and other subantarctic islands), the Grasshopper Moth (*Thyrocopa apatela*; Xyloryctidae, Hawaii), the gelechiid *Ephysteris brachyptera* (Madeira), and the noctuids *Dimorphinoctua cunhaensis* (Tristan da Cunha) and Gough Flightless Moth (*Peridroma goughi*; Gough Island). Some of these extraordinary moths living in such bleak environments are discussed further on p. 152. Interestingly, a few mainland species in which both sexes are flightless live in tunnels and burrows in sandy areas, e.g. the scythridid *Areniscythris brachypteris* (California) and the gelechiid *Catatinagma stenoptera* (Karakum desert). Species with brachypterous or apterous (albeit readily mobile), females do in any case take advantage of their aerially crippled condition in terms of reproductive success. In fact, if a female does not fly, she can save substantial amounts of energy and nutrients that would otherwise be spent building up the massive thoracic flight muscles. These resources can be allocated to the abdomen either to produce more eggs, or larger eggs richer in yolk, to the benefit of the developing embryos.

OTHER WINGED ODDITIES

Moth wings may show several other peculiarities. Sometimes both sexes have patches of raised scales like buttons above fully-reclined scales, such as in Kent Black Arches (*Meganola albula*), Buttoned Snout (*Hypena rostralis*), Gray Eggarlet (*Haplopacha cinerea*) and many species of *Mompha* (Momphidae). Whenever odd patches, tufts or fringes are found only in males, these are usually involved in scent emission during courtship. Scales specialized for scent emission may be directly exposed on the wing surface or margins, such as in the beautiful orange and zebra-striped aganaine erebid *Peridrome orbicularis*, the Scarce Footman (*Eilema complana*), and the large callidulid *Pterothysanus laticilia*, which bears wispy tassels from its hindwings. These scales can also be concealed under protecting folds or pouches, as in males of the erebids *Erebus macrops*, *Hydrillodes toresalis* and *Calesia dasyptera*. Curious lobes borne by the male Seraphim (*Lobophora halterata*), and the Small Seraphim (*Pterapherapteryx sexalata*) are presumably also involved with scent emission. Those of *L. halterata*, at least, conceal hairpencils and some male individuals have much more prominent accessory flaps than others.

A number of species use their wings as sound-emitting organs. Most of these show localized swellings in which a robust, finely ridged section of a vein is flanked by transparent areas. To produce a sound, the vein is scraped by some sort of plectrum, which may be the opposite wing, legs or specialized scales from an overlapping area of the other wing on the same side. Alternatively, the modified region is simply buckled during flight. Adjoining areas work as resonators. These systems are known in many noctuoid moths (e.g. in the genera *Heliocheilus*, *Delgamma*, *Hecatesia*,

BELOW Hummingbird Clearwing (*Hemaris thysbe*; Sphingidae) displaying its abdominal brushes in flight. Most wing scales are lost in the maiden flight.

Pemphigostola, *Hypospila* and *Cossedia*). Sound production in male *Rileyiana fovea* is effected by the scratching of a 'scraper' on the hindleg against a stiff vein at the centre of the hindwing. Oddly enough, that vein protrudes because it lies at bottom of a raised resonating area whose shape resembles a mandolin!

Colour and pattern are of paramount importance in the relationships of a moth with other organisms and the environment, but it is worth emphasizing that many species exhibit transparent areas on wings. Scaleless windows may combine with 'clothed' areas to produce complex patterns such as those seen in the saturniid genera *Attacus*, *Rothschildia* and *Epiphora*, and the wings are almost entirely transparent in numerous day-flying moths (even the saturniid *Heliconisa pagenstecheri*). Best known here are clearwing moths (Sesiidae), wasp moths (Arctiinae: Ctenuchini) and some day-flying bee hawkmoths (*Hemaris*, *Cephonodes*) whose transparency perfects their mimicry of bees or wasps especially once the scales are shed, which happens in the bee hawks during their maiden flights.

LEGS

Moth legs correspond to the original articulated appendages of Arthropoda and as such they consist of a series of jointed parts. They start with a roughly conical coxa partly fused with the thorax, then a small inconspicuous piece called the trochanter which leads to long sections known as the femur and tibia. At the end of the tibia, a many jointed tarsus of normally five short articles (sometimes less) is found. Tipping the tarsus is the pretarsus, a short piece bearing a pair of minute claws and some adhesive pads whose overall function is to grasp and adhere to the substrate. These claws often jab in slightly when you hold a moth on your finger. Legs may show differing degrees of development,

BELOW Hind-, mid- and forelegs of a prominent moth (*Bardaxima lucilinea*; Notodontidae) from one side of the specimen, showing names of the principal segments and structures. The epiphysis is used for cleaning the antennae, while the spurs can help in resting on a surface. Paired claws, a minuscule pad and other microscopic lobes at the tip of the tarsus constitute the 'pretarsus', which improves the grip.

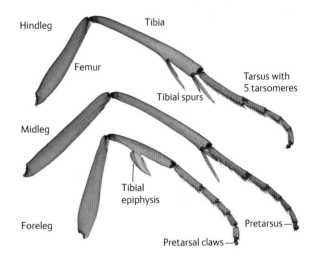

Hindleg
Tibia
Femur
Tibial spurs
Tarsus with 5 tarsomeres
Midleg
Tibial epiphysis
Foreleg
Pretarsus
Pretarsal claws

LEFT The Double Line (*Mythimna turca*; Noctuidae) has conspicuous brushes on its mid- and hindlegs.

ABOVE The female of the Water Veneer (*Acentria ephemerella*) is one of the few moths that can swim underwater, using fringed hairs on its mid- and hindlegs like oars.

modification and scale vesture. On the inner side of the foreleg tibia there is usually a well-developed lobe, named the tibial epiphysis, whose internal surface is comb-like. This structure is used for cleaning the antenna, which is slipped through the groove between the tibia and the epiphysis.

Species pupating underground need to be able to dig themselves out using powerful spines or claws when the moths eclose from their pupae. These claws are found always on the foreleg tibia, but occasionally on the basal segments of the foretarsus too. Smaller spines usually occur along the tarsi and sometimes also on mid-tibiae and hind-tibiae. These two are also often provided with pairs of long spurs, in their most usual arrangement as a pair at the end of the mid-tibia and one just beyond the middle and another one at the end of the hind tibia. Spines and spurs may be used in defence (see p. 107). Legs are also a good location for males to bear tufts or pencils of scent scales. These may occur almost anywhere and on any legs. The mid-legs and hind-legs of the Double Line (*Mythimna turca*) bear conspicuous reddish tufts, whereas those of several species of the erebid genus *Catocala* are concealed in a groove along the mid-tibia. Particularly tufted and modified are the forelegs of many male herminiine erebids, in which the tibia is greatly elongated into a concave cap that accommodates a strongly modified tarsus too, this usually showing a reduced number of articles, sometimes one only, with the most basal being the longest. The mid-legs and hind-legs of female Water Veneers (*Acentria ephemerella*), an aquatic species, are streamlined albeit distinctly bristly, and function like oars for diving.

ABDOMEN

The third main section of a moth's body is the abdomen. This is usually elongated, tubular, and may be viewed as a succession of 7–8 ring-shaped segments followed by two strongly modified ones bearing the copulatory organs. Males show 10 clearly discernible abdominal segments, whereas females of more advanced groups exhibit one less because of a fusion that takes place during metamorphosis. Females of primitive groups such as the Micropterigidae do however still show 10 segments.

A typical abdominal segment consists of two curved plates, a dorsal and ventral one (tergal and sternal, respectively), joined by membranes at its sides (pleurae). The pleurae bear paired spiracles, respiratory openings from which the respiratory tubes (tracheae) branch off into the body. Adjoining segments are connected by a narrow ring of intersegmental membrane. Contrary to the fused segments of the head or thorax most of those of the abdomen can move relative to each other. The first couple of abdominal segments are, however, modified to ensure articulation with the thorax, as are the terminal segments, which are differently configured between the sexes, and can be telescopically retracted or extended. The abdomen is packed with organs and extensive anatomical networks. These include reproductive, excretory, and glandular systems. The abdomen also hosts vital components of the digestive, circulatory, musculatory, respiratory, and nervous systems (see image opposite). Often a large volume is occupied by fat reserves. The abdomen may also carry a wide gamut

David C. Lees and Alberto Zilli
Moths: A Complete Guide to Biology and Behavior
ISBN 978-1-58834-654-4

Errata

p. 7, caption line 3, *for Macrosoma heliconiaria* read *Macrosoma subornata*

p. 11, caption line 4, *for* manoevrability *read* manoeuvrability

p. 12, top caption line 1, *for* hawkmoth (*Oryba kadeni*; Sphingidae) *read* hawkmoth *Oryba kadeni* (Sphingidae)

p. 15, caption line 2, *for* red arrows *read* green arrows

p. 16, bottom caption line 1, *for* glowing eyes *read* glowing eye

p. 45, right hand image is incorrect – correct image is shown below bottom left

p. 81, caption line 4, *for* Six Spot Burnet *read* Narrow-bordered Five-spot Burnet

p. 95, caption last line, *for* misidentified *read* misidentifications

p. 124, caption line 2, *for* ant mimicking a cosmopterigid moth *read* ant-mimicking cosmopterigid moth

p. 128, top caption line 1, *for* Early Thorn (*Selenia dentaria*) read Purple Thorn (*Selenia tetralunaria*)

p. 133, caption line 4, *for* is disturbed *read* if disturbed

p. 141, caption line 1, *for* asket-cocoon *read* basket-cocoon

p. 161, caption line 11, *for* its sister species *read* its sister lineage or species

p. 164, top right hand image is incorrect – correct image is shown below bottom right

p. 184, caption line 3, *for* Paraná *read* Rio de Janeiro; line 5, *add* Even when dying (here attacked by ants), the spines should not be touched.

p. 208, Picture credits, *add* p. 44 (bottom) © Todd Gilligan *and* p. 164 (top right) © Paolo Mazzei

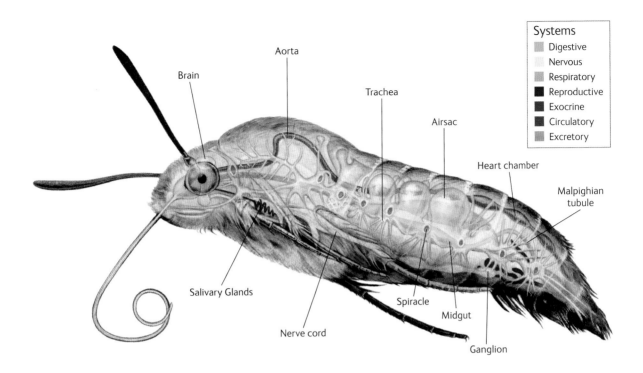

Brain

Aorta

Trachea

Airsac

Heart chamber

Malpighian tubule

Salivary Glands

Spiracle

Midgut

Nerve cord

Ganglion

of hearing organs, sound-producing organs and scent structures. We focus here, though, on the genitalia, which are fundamental in moth systematics and taxonomy.

In males, the genitalia are strongly hardened (sclerotized), as they have to clasp the female parts securely and this rigidity is helped by fusion of 9th+10th segments. The smaller 10th segment is attached to the upper part of the 9th. The 9th segment is particularly conspicuous. This is a stiff ring made by union of the dorsal and ventral plates, consisting of the broad, often hood-like tegumen, and the U- or V-shaped vinculum, respectively. Perhaps the most striking part of the male genitalia is the pair of flap-like valves (valvae) on each side of the 9th segment, hinged to the vinculum. The valves can open and close and often bear elaborate sclerotized processes that are usually distinctive between species. The male intromittent organ (the phallus, more politely termed aedeagus by some moth people) protrudes through a membrane that seals posteriorly the 9th segment. Such a membrane (diaphragm) also exhibits one or more sclerotized regions that support the moving parts, such as the often shield-like juxta, located between the valve bases. The phallus is usually structured as an outer sclerotized shaft and an internal membranous tube (vesica), which is everted into the female genitalia during copulation. At the interior end of the phallus, deeper in the male abdomen, the vesica is linked to the testes through a long membranous duct. The vesica often bears on its surface strongly sclerotized spines or thorns (cornuti), which enhance its adherence to the female parts during sperm transfer. In some groups of moths these spines can be shed in the female, as a 'scorched earth' policy to subsequent males. The 10th segment has two parts. The often hook-like dorsal most part (uncus) is usually present, attached to the tegumen.

ABOVE The anatomy of a Hummingbird Hawk-moth (*Macroglossum stellatarum*), showing the placement through the body of the principal systems (stylized for clarity).

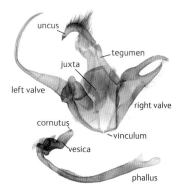

uncus

tegumen

juxta

left valve

right valve

cornutus

vinculum

vesica

phallus

ABOVE The male genitalia of an erebid moth (*Anisoneura papuana*). On top the clasping apparatus with main parts indicated, on bottom the phallus removed from the apparatus to show its configuration. Sometimes male genitalia can be exceedingly asymmetrical such as in this case.

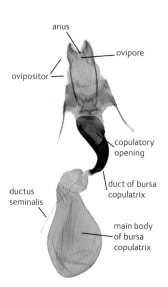

anus

ovipore

ovipositor

copulatory opening

duct of bursa copulatrix

ductus seminalis

main body of bursa copulatrix

ABOVE The female genitalia of another erebid (*Anisoneura aluco*) show the great development of the bursa copulatrix, that often extends well inside into the abdomen.

The ventral component of the segment is the gnathos, paired or fused, which is sometimes missing. In that case, the diaphragm extends up to the uncus. The anus opens into this membranous section of the 10th segment, usually at the tip of a protruding anal tube. The valvae with their processes and the sometimes hinged uncus play the principal role in clasping the female genital segments.

Female genitalia are even more complicated and three quite different configurations are recognized. Most primitive lepidoptera have a single ventral genital opening (the monotrysian type). The vagina thus serves both for copulation and egg-laying. Sperm cells, typically packed into a sort of delivery parcel called the spermatophore, are transferred by the male into a purse-like structure inside the female abdomen called the bursa copulatrix. The spermatophore is delivered through the phallus via the female's vagina. Sperm travel backwards from the bursa to reach the junction where the main oviduct, channelling eggs from the ovaries, is found. Eggs are then fertilized and laid, leaving the mother's body through the vagina itself. Two variations of the theme exist on top of that, i.e. the vagina may be completely separate from the anus so that it opens out independently (as seen in the family Heterobathmiidae), or the vagina and intestine may lead to a common chamber (the cloaca), which opens externally (as in the family Micropterigidae).

In another female type, called exoporian (occuring only in the Hepialoidea), the anus is always independent and there are two genital openings, one used for mating (the ostium bursae) that leads to the bursa copulatrix, and a completely distinct opening connected with the oviduct and ovaries through which eggs are laid (the ovipore). These two openings are typically present on the same abdominal segment (be it the 9th, or the 9th fused with the 10th, as mentioned before). In moths showing this configuration (such as swift moths), however, the sperm has to take a risky journey to successfully fertilize the eggs. Remarkably, sperm leave the bursa copulatrix and reach the exterior of the female abdomen, where they travel along a groove that leads to the ovipore. They then enter the latter opening to be temporarily stored in a small sac (spermatheca) which is a siding adjoining the oviduct. Subsequently, the fertilized eggs descend along the oviduct and will meet sperm cells released by the spermatheca, to be finally laid.

The third and most widespread type of female genitalia is the ditrysian one, where there are two genital openings as in Exoporia, albeit on different segments. Depending on the species, the copulatory orifice can be positioned between the 7th and 8th segment, and the ovipore is always on the 9th (viz. 9th+10th). There is an internal specialized duct for the passage of sperm between the two sections (for copulation and fertilisation) of the genital apparatus. The anus is generally an independent opening, although there are a few exceptions where both the intestine and ovipore open into a cloaca. The spermatophore is as usual transferred and placed by the male in the bursa copulatrix. The sperm then leave the latter and travel along a duct (ductus seminalis) to be transferred to the spermatheca. After fertilisation, the eggs usually receive a sticky secretion from a gland just before oviposition, and thus they are glued to the egg-laying site, usually a plant.

HOW MOTHS GIVE 'BIRTH'

The last couple of segments of females can usually be telescopically retracted within the abdomen, and expanded during copulation and especially oviposition, thanks to two pairs of sclerotized rods for the insertion of retractor and protractor muscles. These rods may be very long and when stowed away at rest, extend far inside the abdomen. The last abdominal segment(s) in fact serve also as an ovipositor and (depending on the biology of each particular species) they show adaptations suited to optimally reach egg-laying sites. The tip of the ovipositor is formed by a pair of setose lobes (papillae anales) encompassing the ovipore and finally the anus. Noteworthy are ovipositors of species that lay eggs into particularly deep sites such as long tubular flowers or crevices in bark and wood. Many species of *Hadena*, for instance, oviposit inside campion flowers (Caryophyllaceae). To reach the inner inflated part (calyx) they have an exceedingly expansible membrane between the last two segments to maximize the ovipositor length. Similar adaptations are seen in many castniids and cossids whose larvae develop within stems and trunks. Many adelids and incurvariids also have an elongated ovipositor whose tip is further modified like a serrated blade, to slice easily into plant tissues. Flattened ovipositor tips are also seen in species that lay eggs tucked into leaf sheaths wrapping the stem of grasses. The same result may be achieved by the papillae anales being compressed either laterally (as for the noctuid genus *Mythimna*) or dorso-ventrally (as in many *Apamea*). The morphology of the egg laying apparatus is predictive of the final oviposition site. Stiff, sclerotized, piercing or cutting papillae anales are always an indication that eggs are syringed into the egg-laying sites, whereas soft ones are usually characteristic of a species that lays its eggs on exposed surfaces such as leaves, stems, bark or rocks. Some species with weakly differentiated ovipositors even lay eggs in flight (e.g. many Hepialidae).

A very different birth strategy, interestingly enough, is (ovo-)viviparity, a phenomenom where live larvae are laid, as for a few species of the tineid genus *Monopis* from the Indo-Australian region. In Lepidoptera the larvae can either hatch within the abdomen or almost as soon as they are laid. Examples are the Viviparous Case-bearer (*Coleophora albella*) and the gelechiid genus *Palumbina*, as discovered by Ga-Eun Lee and Houhun Li in 2017. More needs to be understood about the biology of such remarkable cases, for example whether the new born larvae are nourished by the female in some other way than via the egg yolk, or whether this strategy evolved to evade parasitism.

In this chapter, we have seen reasons why Lepidoptera (mostly moths) have been particularly successful in diversification among terrestrial habitats, and have broken down in detail the complex anatomy of an adult moth that facilitates their spectacular adaptive radiation. In the next one, we see how that beauty and complexity has arisen from specializations of the first three 'early' stages.

ABOVE Oviduct of a Viviparous Moth (*Coleophora albella*; Coleophoridae) from Sardinia, showing the unhatched first instar larvae, with their prominent head capsules, inside.

CHAPTER 2

Becoming a moth

MOTHS NOT ONLY HAVE AN INTRICATELY constructed body but a complex life cycle and before becoming adults (also called 'imagines'), they undergo the three distinct stages of egg, larva (caterpillar) and pupa. Often the importance of these early stages is overlooked; for example, they are poorly represented in museum collections. As a matter of convention, these 'immatures' are not used as 'type specimens', which constitute the fundamental anchor points of moth scientific naming (nomenclature). But the fact is that, with the exception of a few species, the adults of which either hibernate or aestivate, most of a moth's life is spent in these three early stages, known as the pre-imaginal instars.

EGGS

Moths may lay eggs singly, in small clutches or even in masses. The latter strategy may bring some advantages to larvae eating together in terms of better thermoregulation, exploitation of the food resource and protection from enemies (safety in numbers). In some cases, two or more females have been observed laying clutches together. Some of the largest eggs are 4 mm (0.16 in) in diameter and belong to the also impressively large Comet Moth (*Argema mittrei*), though females of a European species, the Oak Hawkmoth (*Marumba quercus*), can oviposit similarly-sized eggs. Surprisingly, there is only a modest correlation between the size of a moth and the size of its eggs. Large species often lay numerous, tiny eggs, painstaking work to count! Some 44,100 eggs were recorded from the abdomen of a single female Red Gum Ghost Moth (*Trictena argentata*; Hepialidae), a giant species that attains a wingspan of 17 cm (6.7 in). That particular female only managed to lay 29,000 of them before death! This is possibly a world record for a member of the family Hepialidae, achieved with a huge abdomen and pinhead-sized 0.6 mm diameter eggs. Even when dried, some Australian hepialids can be so heavy, perhaps in excess of 25 g (1 oz), that a pin used to pick up the specimen is barely able to support the abdomen's weight. In collections, the abdomens of such specimens need shoring up. Giant cossids (Cossidae) such as the goat or carpenter moths of the genus *Endoxyla*, can even be heavier than big hepialids. The Giant Wood Moth (*Endoxyla*

OPPOSITE Pupa of a Costa Rican looper moth (Sterrhinae) secured by a silken girdle. The moth has already begun to take form – compound eyes and marginal wing eyespots are visible.

ABOVE Female of a giant Australian 'micromoth', the Giant Wood Moth (*Endoxyla cinereus*), which is 25 cm (9¾ in) in diameter and weighs around 50 g (1¾ oz).

BELOW A female of the Small Eggar (*Eriogaster lanestris*) which has covered her eggs with a woolly layer of hairs detached from the abdomen.

cinereus; Cossidae), from Australia, is perhaps the largest 'micromoth' (a loose term of convenience) and a single female may span 25 cm (9¾ in) and weigh up to 50 g (>1¾ oz), or the weight of a chicken egg. Female *E. macleayi* are reputed to be even heavier, based on dry specimens. Larvae of such monstrous 'micromoths' are among the legendary 'witchetty grubs' that enrich the diet of aboriginal Australians. Other heavyweights include the stout-bodied cossid *Endoxyla encalypti*, from Australia and Tasmania, and *Duomitus ceramica*, widespread in the Indo-Australian tropics, both of which are reported to lay up to 18,000–21,000 eggs per female, although there is a record of *D. ceramica* laying an estimated 50,000 eggs.

Eggs may be attached to the food source or off it, for example on a nearby rock, or released in flight, like bombs dropped from a plane, where they intermingle with those of other females. As in any of the moth's life stages, the eggs are exposed to adverse weather conditions, predators, pathogens and parasitoids (small organisms acting like parasites that end up in killing their hosts, like predators do). It is no surprise therefore that adaptations have developed to prevent the loss of young caterpillars before they are born. Camouflage helps and eggs are often the same colour as the food source on which they are laid. In addition they are often deposited in sheltered places. Pale and translucent when first laid, the eggs darken into their typical colour after a few days. A first level of defence is represented by the robustness of the egg-shell itself, the chorion. This may be smooth or ornamented by crests and ridges, often intersecting in a way so as to form a finely reticulated pattern on the surface. The shape of the eggs is varied too, including spherical, ovoid, conical and near pyramidal forms. Those in which the micropyle, a small perforated plate where sperm can enter, is on the top are called erect eggs; those in which the micropyle is on a side are said to be reclinate. The egg surface is also finely perforated by aeropyles – minuscule holes permitting gas exchange – either scattered or occurring on particular parts of the shell (often along the elevated ridges). Some moths also protect the eggs once laid. For example, females of the Gypsy Moth (*Lymantria dispar*), Gold Tail (*Euproctis chrysorrhoea*), and Small Eggar (*Eriogaster lanestris*), bear conspicuous tufts of hairs at the tip of the abdomen, which they detach and knead among their overwintering egg clutches to enhance their protection either by insulation or because the hairs may be repellent or obfuscating to predators. Sometimes clutches of eggs may encircle a stem or twig, such as those of lackey moths (*Malacosoma*); the advantage of such neat bangles is unclear.

LIFE STRATEGIES

There is limited correlation between the size of a moth and that of its eggs. Some big moths lay numerous small eggs whereas some small moths may oviposit just a handful of large ones. This leads us to a fairly general (albeit useful) categorization for understanding life strategies of moths, and other organisms as well. As individual reservoirs of resources (energy, nutrients, time) are fixed, a species faces two options to pursue its reproductive success. It can invest in quantity or in quality of progeny. A female may lay a large number of eggs, but these by necessity will have to be small because the fixed amount of available resources has to be shared among them all. Alternatively, the available resources can be used to produce a smaller number of more nutrient-rich eggs. One of the most dangerous periods in a moth's life is that between the hatching of the egg and when the young larva eventually starts feeding and growing, a phase when adverse weather conditions and chances of getting lost without finding the proper food source may decimate offspring even more than predators. Accordingly, freshly hatched caterpillars deriving from low-fed embryos will have limited autonomy before starvation, contrary to well-fed ones. In the first case (*r*-strategy), therefore, the species relies on the number of offspring: many hatchlings will die but due to their high number some will in any case survive and attain maturity. In the second case (K-strategy), each one of the few caterpillars will have greater chances of survival.

There is no definitively better strategy between the two, although the evolutionary and ecological consequences of following one or another may be profound. In fact, *r*-selected species are typically generalist ones, highly mobile and capable of exploiting a wide choice of environments. Due to their great reproductive potential they can sometimes rapidly colonize and exploit sudden ecological opportunities presented to them by nature, but they may similarly suddenly disappear when the fight for survival becomes tougher. In fact, due to their strategy as hit-and-run species, they have less opportunities to finely adapt to the local ecological conditions. Instead, this is a prerogative of K-selected species, more suited for specializing to a local environment. However, if you are strongly adapted to some conditions, it follows that you will not be able to cope with a different set of conditions. Accordingly, K-selected species will show strong specialization for a particular environment, will be capable of exploiting it in a deeply sophisticated manner, but they will be somewhat 'trapped' within it. They would tend therefore to be less mobile, as they have no need to disperse to look for another environment, and they have diminished chances of adapting to a new one anyway. Mature, stable and predictable environments will therefore tend to be filled with K-selected species over time, whereas pioneer, ecologically dynamic and unpredictable habitats will show a higher proportion of *r*-selected species.

Needless to say, we have portrayed the extremes of a continuum of modes of life and many species show a complex array of idiosyncratic strategies and adaptations. They can be widely tolerant with respect to one environmental factor and particularly selective to another one. However, a quick test of their reproductive strategy type when two species are compared is: how many eggs does each of them produce and how large are they?

LEFT The practically wingless female of a Vapourer Moth (*Orgyia antiqua*) does not travel far, if at all, here laying her eggs on the cocoon.

Neal Smith working in Panama took an extraordinary photo of the communal, mass laying behaviour of the Colipato Verde (*Urania fulgens*). Both *Urania* and the related genus *Chrysiridia* from Madagascar undergo periodic mass outbreaks, when the entire foliage of their *Omphalea* hostplants may be decimated by hordes of developing larvae. Contrast this with the opposite case where a female will take great care where she puts a single egg, on or off the plant. One of the authors (DCL) has observed *Urania leilus brasiliensis* laying singly off its hostplant (*Omphalea brasiliensis*) on a spider web in the Brazilian Atlantic forest. In the Madagascan rainforests, the Sunset Moth (*Chrysiridia rhipheus*) sometimes oviposits on grass blades some distance from its hostplant (*Omphalea oppositifolia*), while at other times it lays in batches on the underneath of its leaves. This goes to show that even a single species can have widely differing strategies. Presumably, at critical times this promotes survival against egg predators as well as minute parasitoid wasps. Laying eggs off the plant outwits wasps that use plant volatiles (so-called kairomones) as a clue to locating their moth hosts, thus putting them 'off the scent'.

In the end, to keep a population stable, an average of two offspring must survive to the next generation and phase in the reproductive cycle. However, at times, far more offspring may be produced, and some of these may go on to disperse into other areas and even migrate far away, if indeed they are likely to outstrip local resources. By mid-summer in parts of its range in Europe that it has only recently invaded, the Horse-Chestnut Leaf-miner (*Cameraria ohridella*), will have covered most horse chestnut leaves with its unsightly brown blotch-like mines. Despite the fact that over 60 different parasitoid wasps are known to attack its caterpillars, as well as blue tits that have quickly learned to rip open the leaf mines to get at the grubs inside, at each generation (there can be three in a season) the leaf-miner population shows an exponential growth pattern. The larvae eventually crowd each other out on the leaves and even do not seem to mind sharing the same leaf mine (see p. 158). At times so many adults emerge that there is a veritable 'aerial plankton' of individuals which disperse to colonize new areas.

CATERPILLARS OR 'HAIRY CATS'

Hatching is often signalled by a change of the egg colour. Emerged caterpillars may wholly or partially eat their empty eggshells and sometimes they do not eat any part. By eating the eggshells they assimilate nutrients that would otherwise be lost, but in doing this they waste time that could be spent starting to feed on leaves. Caterpillars of some species can take on viral or bacterial infections from the eggshell and die. Entomologists have discovered that larval health can be improved if eggs are sterilized with a very weak bleach solution.

The word 'caterpillar' derives from the Latin '*catta pilosa*', meaning 'hairy cat' (see p. 138). Lepidoptera larvae are said to be 'eruciform' (since *eruca* is the Latin for caterpillar), a type shared by a few other insect groups, such as the scorpion flies and the sawflies. A typical eruciform larva is elongated, cylindrical and soft-bodied, with chewing apparatus, comparatively small legs on thorax (the true legs), and a variable number of additional fleshy 'legs' (prolegs) along the abdomen. In a caterpillar's body, we can easily recognize the head and trunk. In the latter, the distinction between the thorax and abdomen is not immediately evident as segments are essentially identical, but the three thoracic segments, immediately behind the head, can easily be spotted by the presence of a pair of true legs on each. These are short, roughly conical in shape, but curved inwards, and articulated into short segments shielded by hardened cuticle and terminating in a hooked claw.

In the abdomen 10 segments are readily discernible, while remnants of an 11th segment possibly manifest as soft lobes surrounding the anus. In the most typical configuration, prolegs occur from the 3rd to the 6th abdominal segments and on the 10th, but there are many possible deviations from this arrangement. Perhaps best known are caterpillars of the 'loopers' (Geometridae), which have prolegs only on the 6th and 10th abdominal segments. Loopers travel by stretching the front part of the body, grasping the hostplant with the true legs, and then pulling up the abdominal segments nearest to the thorax in a looping action. The 'semiloopers' (some Erebidae and Noctuidae) lack prolegs on the 3rd segment (and often the 4th too) but still loop in this fashion. Some caterpillars, such as the White Ermine (*Spilosoma lubricipeda*)tend to 'trundle' along – the prolegs are lifted, each pair in turn, another method which enables very fast walking. In some families (such as the Heliozelidae), caterpillars have completely lost prolegs or just show rudiments, particularly in types living in concealed sites such as leafminers burrowing inside leaf blades.

A typical proleg consists of a fleshy ventral wart protruding from the body that bears on its tip a soft extrusible pad (the planta), with one or more series of minute hooks (crochets),

BELOW Caterpillar of the Banded Woolly Bear (*Pyrrharctia isabella*) showing its tightly gripping prolegs. Woollybear festivals are held in several locations of USA in the fall.

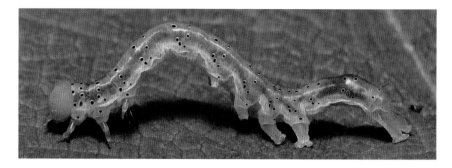

whose function is to ensure a firm grasp to the hostplant. Some caterpillar biologists indulge in a sort of 'crochetology' with a series of classifications of crochet arrangements across different families. In some groups, the most posterior pair of prolegs has lost any walking function, the legs having been modified into protuberances to enhance camouflage, for example in the Rose Hooktip Moth (*Oreta rosea*), or to offer defence, such as in the Puss Moth (*Cerura vinula*). Slug caterpillars (in the family Limacodidae), have a greater number of small, sucker-like prolegs devoid of crochets, with which they essentially glide rather than walk or crawl.

BUILT FOR CHOMPING AND CHEWING

A caterpillar's head is profoundly different from that of an adult. Its orientation with respect to the body may vary from having the mouthparts facing forwards, such as in leafmining or wood-boring forms, to those more or less facing downwards, as in most leaf-eating species. Most of a caterpillar's head is taken up by paired lateral caps which meet on top of a triangular frontal plate. They bear simple organs of vision, the stemmata, usually six in number, which appear on the cuticle as small lenses. Their visual ability bears no comparison with the compound eyes of the adults, nonetheless there is evidence that they produce coarse images. They are in any case sensitive to colours and changes of the light conditions, informing the caterpillar about day or night and therefore the appropriate times to rest and hide. The larval antennae are short, usually articulated into three segments and are richly provided with sensory bristles and a handful of sensilla. The mouthparts are basically structured like those of other chewing insects, with a superior transverse labrum plate, paired mandibles, paired maxillae and an inferior labium, although there are several peculiarities in the configuration of maxillae and labium. Mandibles, which do most of the chewing, can be asymmetric and may protrude only with their tips so they do not always appear particularly big. The broad flat labium extends into the bottom of the mouth cavity and fuses with a fleshy albeit spiny lobe helping the swallowing of food (the hypopharynx), while on the exposed ventral side, the labium ends in short paired labial palpi and a spindle-shaped protrusion at its middle, the

BELOW Head of Tobacco Budworm (*Chloridea virescens*) caterpillar with main parts labelled (mandibles and labio-hypopharyngeal complex are shown separately, slightly enlarged). Nomenclature of setae is not shown.

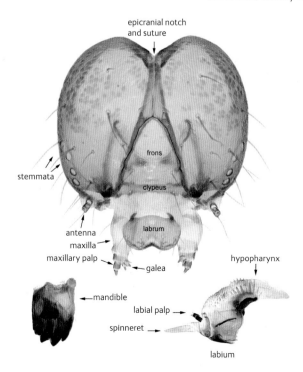

epicranial notch
and suture

frons

stemmata

clypeus

labrum

antenna

maxilla

maxillary palp

galea

hypopharynx

mandible

labial palp

spinneret

labium

spinneret. This specialized structure is the nozzle through which silk passes, secreted by a pair of glands. The ability to produce silk is widespread across the Lepidoptera and some other insect orders, but it is in cocoon-spinning species that it achieves its pinnacle of artisanry. Suffice to say that in the Silkworm (*Bombyx mori*) and many other silk-producing species, the silk glands are so developed and convoluted that they extend well into the caterpillar's abdomen. The spinneret is absent in some primitive moth families like Micropterigidae and many internal feeders. A larva with a spinneret was found in Lebanese amber (about 125 million years old) – the organ is an evolutionary novelty of those Lepidoptera that exhibit a proboscis (the so-called Glossata).

HISSERS AND STINGERS

Caterpillars exhibit several other curious features. Easy to spot due to colour contrast are the respiratory spiracles running like small portholes along each body flank. These occur on the first thoracic segment and along the abdomen from the first to the eighth segment. The entire caterpillar body also bears sensory 'bristles' known as setae. Entomologists sometimes avoid using the term hairs (which strictly speaking are a feature of mammals), but in this book we use the term loosely, for convenience, particularly to refer to secondary setae, i.e. those appearing in a caterpillar after its first moult, or the thin, hair-like scales which, along with flat ones, clothe the adults. The academic study (known as chaetotaxy) of both the primary and secondary setae, which a caterpillar is born with, has great importance in tracing the evolutionary relationships among moth groups. The precise pattern formed by these sensory setae reflects the distribution of underlying nerve pathways, and particular setae can be given a standardized code across the Lepidoptera so that researchers can compare the setal pattern of two or more species to get an idea of their affinities. Such setae are best viewed down a microscope on a newly hatched larva and have nothing to do with the often conspicuous and long 'hairs' (i.e. secondary setae) that cover the body of many caterpillars. In fact, caterpillars may either have a smooth surface with its

BELOW LEFT The Squeaking Silkmoth (*Rhodinia fugax*) caterpillar expels air through its spiracles (here white, two are arrowed) making a strange and presumably snake-like hissing noise, as does the Walnut Sphinx (*Amorpha juglandis*). Other caterpilars make a sound evoking the general alarm call used by many birds.

BELOW A 'Red Devil' slug moth (limacodid) caterpillar from Yunnan, China belonging to the genus *Setora*. Slug moth caterpillars are like the stinging nettles of the insect world. To advertise their stinging spines, many species have very striking warning colours.

own coloured pattern, or be variously provided with hairs, lobes and protuberances. Hairs may clothe the body more or less uniformly or they may be organized into bundles and tufts, which sometimes resemble old-fashioned shaving brushes as in the Vapourer (*Orgyia antiqua*) and the aptly named Pale Tussock (*Calliteara pudibunda*).

Many larvae do not have a classical cylindrical body shape. Some bear fleshy knobs on one or more of the trunk segments (e.g. the Looper Moth (*Apochima flabellaria*)), others long thin processes (e.g. the Owl Moth (*Brahmaea wallichii*)) or conical protuberances, which can either be smooth or thickly covered with branches or spines. In some groups, these spines may be fiercingly armed and stinging (e.g. the American saddleback caterpillar genus *Acharia*, and the saturniid genera *Automeris*, *Dirphia* and *Hylesia*). Even the loose hairs of some species can be strongly irritant, notorious among which are those of processionary moths (Notodontidae: Thaumetopoeinae).

Most caterpillars in the family Sphingidae (hawkmoths), but also some in the

ABOVE RIGHT The Monkey Slug (*Phobetron hipparchia*), the astonishing larva of a hag moth with its weird uneven horizontal projections.

RIGHT An apatelodid shag-carpet caterpillar (*Prothysana felderi*) from Ecuador, showing really unusual scale-like hairs amongst its 'shag pile'.

Bombycidae, Endromidae, Brahmaeidae and Notodontidae, carry variations of a posterior horn on the 8th abdominal segment. The larvae of the hawkmoth genus *Rhagastis* in Asia also have eyespots on humps of the second abdominal segment, but in *Rhagastis lambertoni* from Madagascar (see p. 119), these raised projections have an extraordinary development into short conical tentacles. In many cases, particular segments have undergone variation in size and shape so that the larva has an unusual appearance. Noteworthy is the caterpillar of the Lobster Moth (*Stauropus fagi*), which especially when freshly hatched, appears to be an ant-mimic (p. 125). Caterpillars in the family Notodontidae have such bizarre shapes and postures that they are very difficult to spot at rest (e.g. the prominents *Harpyia milhauseri* and *Notodonta dromedarius*). A somewhat opposite trend is that shown by many slug moths, in which the obvious segmentation has been partly obliterated so that their larvae appear to be dorsally covered by a rather uniform shield (e.g. in the Festoon, *Apoda limacodes*).

ELASTIC SUITS FOR GROWTH

A caterpillar's life is split into several instars. Despite all the advantages of having a bodysuit composed of a chitinous cuticle, this 'fabric' cannot be indefinitely enlarged. It is unable to keep up with the prodigious growth rate of the larval body, which from egg-hatching to pupation may easily undergo a mass increase of several thousand times (e.g. in the range of 7,000–10,000 times in the Silkworm). Regular substitution with a new 'skin' is then required until the limits of that semi-elastic 'jacket' is reached, which is particularly true for the head capsule, invariably harder than the cuticle lining the trunk. This process of skin changing is known as moulting (moult or ecdysis are the technical terms for this phenomenon). Ecdysis is one of the riskiest times for a caterpillar. In fact, the larva has to both firmly grasp a leaf of a hostplant and immobilize itself, in order to secrete a fresh soft cuticle under the

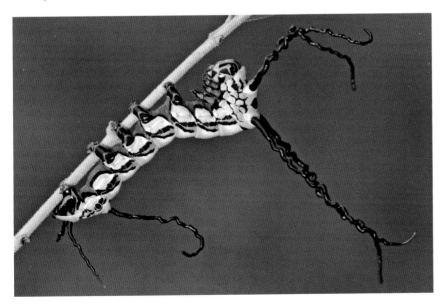

LEFT An Owl Moth caterpillar (*Brahmaea wallichii*) from the Himalayan foothills has bizarrely shaped protruberances, looking like an organism from another planet.

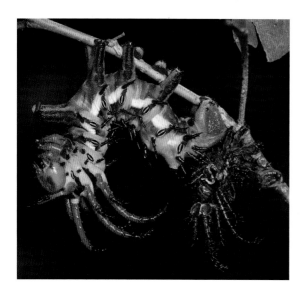

ABOVE The caterpillar of the Hickory Horned Devil (*Citheronia regalis*) after shedding its 'skin'. Moulting is not always straightforward in caterpillars as it depends on the complexity of their form.

old one, and then detach and slip off this old skin, which is cast off (if not eaten) as a thin flimsy larval exuvia. During this operation, which usually takes two to four days, the larva is exposed to predators, weather and even chance events that might cause it to fall from its perch or blow away. It is advantageous for the larva to remain motionless, otherwise the delicate physiological process of moulting might go wrong and the caterpillar would be trapped and choked with pieces of the old, inextensible cuticle. The number of moults that a caterpillar undergoes is quite variable (normally there are four to five instars) depending on the species and its particular life strategies; environmental variables play a part, too. In some species females undergo one extra instar allowing them to grow bigger. There are exceptions; some Lepidoptera have been recorded as having 14 or more instars.

Examples are glass moths (Dalceridae) and the Arctic Woolly Bear (*Gynaephora groenlandica*), which is known to have up to 14 instars, each corresponding to just a year's cycle (one moult), so the larva (see p. 150) can live for 14 years. This extraordinary caterpillar uses a kind of antifreeze in its haemolymph to resist temperatures about as cold as it can get (-70°C). Meanwhile the Common Clothes Moth (*Tineola bisselliella*) can have as much as 16 instars. Similar feats have been achieved by the Goat Moth (*Cossus cossus*) over three years, while the Cottonwood Clearwing (*Paranthrene dollii*) and the Stalk Borer (*Papaipema nebris*) can notch up an impressive 17 instars. Even more may be possible in such psychids as *Oiketicus kirbyi*, although those bagworms typically have nine instars.

In some species caterpillars may dramatically change their appearance between instars. Leaf-mining larvae are an exceptional example of this phenomenon (known as hypermetamorphosis). Early instars of gracillariids have a very flattened, legless, sap-sucking stage whereas later fully legged instars are more cylindrical and feed on tissues. Ectoparasitic Epipyropidae are very diverse in their feeding biology, sometimes moving from one host (e.g. leafhoppers, see p. 60) to grazing on cicadas or other bugs accompanied by a great change in shape. Another ectoparasitic family that shows a dramatic change in larval body form through its life cycle is the Cyclotornidae (see p. 68).

A caterpillar may strongly differ in colouration between instars, as well as occasionally also in shape. The Alder Moth (*Acronicta alni*), whose fully grown larva is black with conspicuous yellow dorsal patches, instead in the early instars has an off-white part and a brownish-black part variegated with white, which combined with a glossy shine and its J-posture, makes it strikingly resemble a fresh bird-dropping. These examples of age-related variation in caterpillars, however, always follow a fixed scheme of development for the relevant species and should not be confused with colour and pattern polymorphisms shown by some species (nor with stronger forms of hypermetamorphosis). In fact, polymorphisms occur regardless of age, so that larvae of the same instar may conspicuously differ, such as those of the Death's-Head Hawk-moth (*Acherontia atropos*), which may be green or yellow with oblique stripes, or speckled with various hues of brown and with a broad white patch on the thorax. Usually, polymorphisms are genetically fixed, that is given a particular gene (or combination of genes), this will forever produce the same pattern irrespective of the conditions experienced by the developing larvae. However, different colour (and shape) variants may also be environmentally determined. This happens when it is the conditions which the organisms face during development that modulates their appearance (polyphenism). In this case, even genetically alike individuals will be differently looking if they are grown under different conditions. It was shown that in some heliothine noctuids, and hawkmoths such as *Daphnis nerii*, larval colour polyphenism depends on the parts of the hostplants the caterpillars feed on. This is to improve their camouflage. Accordingly, green or reddish-brown colour phases develop depending on whether they feed on green parts of a plant or its flowers.

A different, and extreme, case of polyphenism is the seasonal one exhibited by the North American looper *Nemoria arizonaria*, a species with two generations per year feeding on oaks. Caterpillars of the first generation feed in early spring and encounter catkins more often than leaves. They have a yellowish-ochre variegated colouring while their shape closely matches that of the catkins, achieved with flanges and other protuberances from the body. In contrast, the caterpillars of

BELOW The larval instars of the Giant Peacock Moth (*Saturnia pyri*) are strikingly different, turning from black (below) to green (below left and bottom left), and with tubercles changing from orange/yellow to turquoise blue, and yet the caterpillars can blend well into vegetation in later instars.

ABOVE Larvae of the emerald moth *Nemoria arizonaria* radically change shape according to the seasonal brood, eating oak catkins in spring (top) or leaves in summer (bottom), when they blend into the stem.

the summer generation rely on leaves and their bodies are much smoother and coloured a brownish-grey like the twigs they rest upon.

In a number of moth families the young larvae spin a portable case from which they protrude during feeding and walking (see p. 105). They use the case as a retreat if in danger or when resting. These larval cases are regularly enlarged as the larva grows and will eventually be used as cocoons. Particularly elaborate and often bizarre in shape may be those of moths in the families Psychidae and Coleophoridae, though amazing cases occur also among the incurvariids, adelids, tineids, xyloryctids, oecophorids, and mimallonids.

THE PUPA – A CASE FOR CHANGE

'An Aurelia is born of a Caterpillar indeed, then from that a Butterfly'. This is how Ulysses Aldrovandi summarized in 1602 the metamorphosis of Lepidoptera. But what is an Aurelia? Pupa, aurelia and chrysalis are all terms relating to the last pre-imaginal instar in the life cycle of Lepidoptera. The pupal stage is bound by two moults, one when the mature caterpillar undergoes its last skin change to become a pupa, and another one when the pupa breaks to release the fully developed adult (known as the emergence or eclosion of the imago). Under the cuticle of the often motionless pupa, dramatic changes are taking place. In addition to the complete restructuring of the larval body into that of an adult, quite often the so-called pharate (fully formed) adults remain embalmed (even over the winter in *Orthosia*) under their pupal cuticles waiting for proper external stimuli. Usually the cue is a specific external temperature or change in daylight hours (or a combination of the two), which tells the moth the optimum time to emerge. Strictly speaking, aurelia and chrysalis refer to butterfly pupae. In fact, naturalists of old noted that the pupae of many butterflies had shiny golden spots and named them with reference to gold (*aurum* in Latin and *chrysós* in Greek). This is why many naturalists prefer to use the term pupa for moths, while for butterfly pupae the more specialized term chrysalis is often used.

During the pupal stage a massive transformation of the insect from the larval to the adult stage is achieved. Before moulting into a pupa, however, caterpillars undergo a series of physiological changes and enter a phase known as the prepupa, during which their body appears to be contracted and they are relatively motionless. Generally the pre-pupal period begins after the caterpillar's last meal and ends when the larval cuticle is shed from the pupa, when they may turn darker. The activation of this process is triggered by a sharp decrease in juvenile hormone, which had so far acted to maintain the insect in the larval stage. Another hormone, ecdysone, whose relative balance with the juvenile one governs the cycle of moults between larval instars, thus prevails and leads the mature larva to pupate.

Moth pupae appear as roughly fusiform (spindle-shaped), with generally brown or blackish cases in which parts of the developing adult are finely engraved on their surface (see p. 38). On the head and thorax these engravings represent the boundaries of lobes and small sheaths containing the various body parts, which are closely pressed into the pupal surface, whereas on the abdomen they mark the different segments, sometimes with constrictions too. Usually lobes corresponding to eyes, mouthparts (especially proboscis), wings and legs can easily be discerned, and species in which any of these pieces is particularly developed in the adult would usually have a correspondingly larger case in the pupa. This occurs in many sphingids which have such a long proboscis that the case usually protrudes out of the pupa. It also occurs in the antennae of male longhorns (Adelidae) where the thin cases often exceed the pupal length and are rolled around the abdomen (see p. 14). Large mandible cases are present in the pupae of Eriocraniidae. In general pupae of primitive Lepidoptera bear appendages in sheaths comparatively free from the body, while in more recently evolved groups they are closely fused with the main pupal case (so-called Obtectomera, see p. 4).

Pupation may take place directly exposed to the environment, sheltering under bark, stones, in rolled leaves, in the ground, or in a specially tailored silken bag spun by the larva, the cocoon. Although relatively capable of motility of body parts, especially so in less advanced Lepidoptera, the pupa is almost as inactive a life stage as is the egg. However, a few species seem to have subterranean pupae that are quite prone to locomotion, as Enrico Stella, a breeder of *Brahmaea europaea*, has noted: "they could never be found in the same place!". This ability relies on grabbing the soil with hirsute strips of spines and jerky movements of the abdomen. In such pupae, subterrannean (or below bark) mobility is essential to aid the eventual eclosion of the adult at the surface.

Cocoons are usually hidden in hard to find places and provide additional protection

ABOVE The tail end of a pupa of *Striglina* sp. (Thyrididae) showing elegant cremaster hooks, processes which anchor the pupa in place to its silken attachment (e.g. cocoon), thus facilitating emergence of the adult.

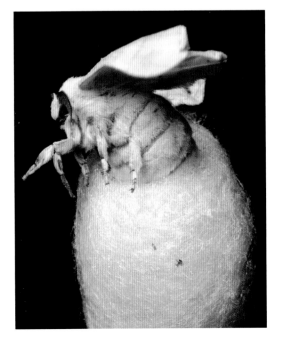

FAR LEFT A silkmoth (*Bombyx mori*) already starting to expand its wings as it struggles out of its cocoon.

LEFT The pupa of the Currant Clearwing (*Synanthedon tipuliformis*) has a 'beak' to help it break through the 'cap' prepared by the larva for its exit.

to the moth for the last period of its metamorphosis. The main spinning fabric, silk, is a remarkable fibre, used for protection in a wide variety of ways. It is composed of minute fibrils of the protein fibroin bundled in another sericin. Sometimes cocoons are very tough and provide a relatively isolated chamber from the exterior to the developing organism (e.g. *Saturnia pyri*), others are so loose and made with so few threads that their only function is to anchor the pupa and prevent it from falling down (e.g. the Ash Bud Moth, *Prays fraxinella*). The distinctly reticulated cocoons of the handsomely red and white striped Muslin moth genus *Cyana* are remarkable in that their mesh-like structure is formed by hairs of the larva bent and enveloped with silk threads. The pupae are kept suspended inside these cocoons on a delicate silk hammock, probably to avoid direct contact with the hostplant and the spread of mould. The practice of mixing larval hairs with the cocoon silk is common in many arctiine and lymantriine erebids and some cocoons can be quite urticating. Amazingly reticulated cocoons are also those of the suspended type exemplified *par excellence* by members of the family Urodidae (see p. 105). Other groups use threads to clog together earth particles or use plant debris to create a protective chamber.

Silk is not only a useful invention for protecting pupae in the cocoon, as some caterpillars spin large communal nests to which they return after foraging. Caterpillars that do this include species of ermine moths *Yponomeuta*, and the eggar moth genera *Malacosoma* and *Eriogaster*, as well as the well-known processionary moths (Notodontidae: Thaumetopoeinae). These nests provide shelter, warmth and protection from enemies and they smooth out weather fluctuations. Silk trails impregnated with a pheromone soliciting a 'follow-the-leader' behaviour have been shown to stimulate communal feeding in *Malacosoma* caterpillars, particularly useful when larvae are small and have to forage on tough leaves.

BELOW LEFT Communal larval nest of the Eastern Tent Caterpillar (*Malacosoma americanum*). The caterpillars exit from the safety and insulation of their 'tent' to eat leaves together. The tent may be 10˚ above ambient temperature in early spring, accelerating metabolism and growth of the inhabitants.

BELOW RIGHT The finished cocoon of an Oak Eggar (*Lasiocampa quercus*).

ACHIEVING THE RIGHT BALANCE OF LIFE

Duration of the various life stages that moths undergo during their life cycle depends on the particular biological strategies evolved by each species. Even within the same species, completion of the life cycle may require quite different timespans depending on whether individuals will be developing during favourable seasons or not. For instance, some species show only one generation per year, but others are multi-brooded and produce two or more annual generations. The concept of voltinism specifies the number of generations by a species over a year. There are species that are univoltine (monovoltine), bivoltine, trivoltine… or multivoltine (polyvoltine). Where seasonality is marked, such as in temperate regions, not surprisingly generations taking place during favourable seasons attain the adult stage more quickly than those experiencing harsher climatic conditions. The well-known Death's-head Hawk-moth (*Acherontia atropos*), is bivoltine in the Mediterranean region, but the time to develop taken by its second annual generation, in late spring-summer, is much shorter than for the first generation, whose larvae grow and undergo pupation in autumn, while pupae remain dormant all winter long and adults are eventually on the wing next spring. When organisms enter phases of dormancy or quiescence during their life cycle, usually to overcome unfavourable seasons such as winter in the example above (hibernation), these are known as periods of diapause, a hormonally induced slowing-down of vital functions.

Unfavourable and favourable are, however, relative terms, and if it is generally true, at higher latitudes, that winter is a disadvantageous season for many moths, this does not hold for the many winter-flying species. These take advantage of being active during the least crowded period of the year, e.g. chestnuts (*Conistra* spp. Noctuidae) and genera of loopers with brachypterous or wingless females such as *Agriopis*, *Erannis* and *Operophtera*. The same applies to species that instead undergo the summer equivalent of hibernation (aestivation), like the Australian Bogong Moth

LEFT Vast numbers of Bogong Moth (*Agrotis infusa*) mass together after migrating from the arid plains to aestivate in caves in the Australian Alps at the start of the summer.

(*Agrotis infusa*) as a means to overcome summer drought. Diapausing stages in Lepidoptera can vary. For example, most species of tent caterpillars (*Malacosoma* spp.) overwinter as eggs, though strictly speaking they do so as larvae that remain inside the eggs until hatching takes place. Larval diapause is known in many groups, above all arctiine erebids, and in lasiocampids like the Fox Moth (*Macrothylacia rubi*). Overwintering larvae of the Heterogynidae and many Zygaenidae also spin and hide inside a temporary cocoon ('hibernaculum'), where they wait until spring. Examples of diapausing adults are those of the Herald (*Scoliopteryx libatrix*) and other erebids like *Apopestes spectrum* and many species of snouts (*Hypena* spp.) which all overwinter in caves and other shelters. However, the most common diapausing stage in moths is the pupal one, either to get through cold periods as for species living in cool and temperate areas, or to await seasonal rains as for those living in desert regions.

The Death's-head Hawk-moth is also interesting as it shows a geographically varying voltinism. In warmer and less seasonal southern parts of its range, the African tropics, it is continuously brooded, and has no diapause, producing up to 10 generations per year. At the other extreme, if environmental conditions are not favourable enough to allow at least one generation per year, other species respond by becoming less than univoltine ('semivoltine'), and take more than one year to accomplish their cycle. This is the rule for several moths occurring in very cold environments, such as the subarctic tundra or at high elevation on mountains. These moths usually have a biennial cycle, which implies that in a given area the adults of a particular brood will appear every second year. Two populations of a same species may thus regularly alternate in such a place. Semivoltinism also occurs when weather conditions are relatively favourable and constant but the food needed for growth and development is poor in nutrients or tough work to eat. Caterpillars of many species live in protected micro-environments that mitigate climatic fluctuations and provide them shelter such as inside stems, twigs, seed-capsules, under bark or in tree logs. Wood-boring larvae, in particular, need to eat a great deal of wood as the nutritive value of this ligneous matter is particularly low, necessitating extra instars, probably also requiring intervention of symbiotic micro-organisms capable of degrading lignin.

As we have just seen, there is room for several modulations and adjustments of the number of generations and relative duration of the various life stages in relation to external conditions. However, sometimes the stages follow a stable pattern. This occurs when the phases in an insect's life are triggered by precisely recurrent stimuli such as when certain daylengths are attained, which happens on a strict timescale at a given latitude. A strict photoperiodism like this, however, is quite

BELOW The larva of the Goat Moth (*Cossus cossus*) has to bite powerfully through solid growing wood and its development there may span 2–5 years, as going is tough as well as nutrients hard to come by.

rare, and organisms usually respond to the interaction between fixed spans of daylength and more fluctuating cues such as temperature and moisture. However, as the relative susceptibility of every species to one or other of the two components is variable, this is what determines whether they will have more or less stable or uneven life cycles.

Moths occurring in strongly seasonal, albeit predictable, environments will have spent several thousand generations finely adapting to the seasonally varying ecological factors they will likely keep on experiencing in the years to come. In contrast, haphazardly variable environments are unevenly exploited by generalist wanderer species (r-selected, see p. 41) or by organisms capable of modulating their life cycle to cope with such eccentricity. In addition, if environmental conditions deteriorate, numerous species react by entering prolonged, even multiannual diapause. A record holder in this respect seems to be the pupae of one of the Yucca moths (Prodoxus y-inversus) which, kept under artificial conditions, are known to have emerged after 16 to 30 years of diapause. Pupae that diapause for five or six years are common in the eggar moth genus Eriogaster, with another exceptional case of a moth emerging after 14 years recorded for the Small Eggar (Eriogaster lanestris). Periods of up to nine years of pupal dormancy have been recorded in zygaenids, sphingids, saturniids and notodontids.

Two aspects have to be stressed with respect to prolonged quiescence. First, this does not necessarily indicate that environmental conditions remain unsuitable. End of dormancy is usually preceded by a sensitive period during which the organism analyses surrounding conditions via the stimuli that it perceives and only if these surpass given thresholds does the physiological termination of diapause begin. Thus, if during such a timespan the environmental cues do not attain the thresholds for activating the response, the insect would carry over the diapause until a new sensitive period begins, maybe for the next year too, irrespective of whether the external conditions become suitable or not. Second, as sensitivity to environmental changes is genetically variable, there can be differences between populations and also among siblings, so that a brood of brothers may differently react to the same external conditions and eventually not be synchronized anymore. Both the capacity of modulating the life cycle and the unevenness of responses to the same stimuli bring another set of adaptive advantages. They may well disrupt, at least in part, synchronization of parasitoids to the life cycle of their hosts and maintain a 'reserve stock' of individuals to overcome unexpected climatic extremes. Were all the individuals of a moth to emerge at once, the entire population would be at risk of succumbing to a sudden deterioration of weather conditions. Hedging your bets is the rule in desert regions with variable water availability and hostplant condition.

Here we have dealt with the basic strategies used to deal with the challenges of survival that are involved in the early stages of moths, together with some of the more interesting behavioural and morphological modifications that accompany them. In the next chapter, we hone in on the extraordinary diversity of approaches moths take to feeding in just two of their life stages.

ABOVE A perfect bilateral gynandromorph of Comet Moth (Argema mittrei) in which the left side is male and the right female, with shorter and wider tails. During embryonic development, mistakes in how sex determining factors are passed on to and instruct dividing cells affects some of them to develop as one sex or the other. Derivatives of these cells of the developing individual will produce feminine or masculine tissues. Usually this leads to an uneven mosaic of sexes in the resulting moth, but if the mistake happens at the earliest phase of embryonic development, extremely rare individuals like this one can be produced, which are unable to reproduce.

CHAPTER 3

A matter of taste

I N THE PREVIOUS CHAPTER, WE DESCRIBED the moth's four-stage life cycle. Two of the stages (egg and pupa) are permanently non-feeding, whereas the other two stages – caterpillar and adult – are those during which feeding can take place. Caterpillars are particularly devoted to eating, while adult moths must balance the need to feed with reproduction and dispersal, and many species with reduced mouthparts do not feed at all. There are many different feeding strategies and food sources used by moths. Here, we examine some of the more bizarre dining habits of moths, either as caterpillars or as adults.

CATERPILLAR MEALS

Moth caterpillars have one overarching function: to eat, but what they eat can differ markedly. The majority munch on leaves or other plant parts. Mealtime for others is extraordinary, for example devouring antelope horns, sloth poo and tortoise shells. Some share our homes and warehouses, chewing on woollens and stored food. Exceptional also are carnivorous larvae, among them those that have evolved some ingenious ways of not being eaten themselves as they go into the 'lion's den', hiding in plain sight in ants' nests.

CARNIVORES TO MATRIVORES

Bagworms (Psychidae) are related to clothes moths and other tineids (Tineidae). They have a broad diet and voracious appetite. The giant bagworms in the genus *Deborrea* from Madagascar can strip a wide range of trees completely bare, even when the caterpillars are infested by tachinid fly parasitoids. Another bagworm species from Panama, *Perisceptis carnivora*, is predatory on a wide range of arthropods, including spiders and grasshoppers. Macabre as it sounds it decorates its larval case with the remains of its meals.

It has long been suspected that some bagworm moths have a more gruesome diet at the start of their life, with hatchlings being carnivorous over their motionless mothers when they are all in their pupal cases. Matrivorous habits are definitely the norm for species of the genus *Heterogynis* (Heterogynidae), whose larvae entirely

OPPOSITE A Yucca Moth (*Tegeticula maculata*) from California which has a symbiotic relationship with a Yucca (*Hesperoyucca whipplei*). It is the yucca's sole pollinator and in turn the caterpillar is allowed to feed on some of the seeds. Notice the large pollen load that the moth clutches with its peculiar tentacle organs, here congealed in a museum specimen.

devour their mothers soon after hatching. Female *Heterogynis* accumulate fatty deposits in their bodies, providing their progeny with a super-rich first meal before the young caterpillars disperse onto the hostplants. After protruding from their cocoons to attract and copulate with the males, which are winged and able to fly, the wingless females come back and oviposit into their pupal exuviae. Then they enter a motionless 'subvital' stage effectively preserving themselves as fresh meat for the offspring until the eggs hatch, some 10–12 days later. A similar strategy of boosting the offspring's chances of survival with an initial energy-meal has been observed by star of the Peppered Moth story, H.B.D. Kettlewell, in the African vapourer moth (*Bracharoa dregei*; Erebidae). The young of another erebid, the Andean arctiine moth *Andesobia jelskii*, also initially feed off the dead or dying mother. The Lepidoptera are usually thought to be a group of animals that do not exhibit parental care. However, is it not the ultimate sacrifice for a mother to pass on her body to that of her children?

DEVOURING OTHER CATERPILLARS, LIVE FLIES AND SNAILS
It has long been known that some caterpillars like the Dunbar (*Cosmia trapezina*), are cannibals, devouring other caterpillars of their own species as well as others. It is an initially mystifying rearing experience, to find the number of Dunbar caterpillars in a box rapidly dwindle to one. Yet truly predatory caterpillars are rare. Hawaii has some of the most remarkable instances known of carnivorous moths. At least six, and maybe most of the 22 endemic species of Hawaiian pug moths (including *Eupithecia orichloris, E. scoriodes* and *E. staurophragma;* Geometridae) feed on flies or other insects. The caterpillars mimic twigs at rest, but they have sensitive bristles which, when triggered, cause the caterpillar to suddenly snap backwards and snatch flies (though not literally out of the air) with their long, spiny, grasping legs. Anal projections may serve a similar function, acting like trip wires, so that the whole caterpillar is rather like a Venus Fly Trap. This pug moth behaviour was an amazing discovery in

RIGHT A caterpillar of a carnivorous pug moth from Hawaii (*Eupithecia orichloris*) rests motionless, its twig-like body firmly anchored by the enlarged hind prolegs, until the moment that its sensory hairs trigger it to snatch any insect that approaches close, using its extended, raptorial true legs.

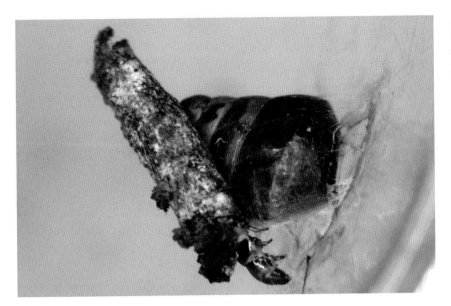

The larva of the appropriately named *Hyposmocoma molluscivora*, another denizen of Hawaii, takes carnivory to another level by directly feeding on a live snail.

1983 by Hawaiian biologist Stephen Montgomery. Pug larvae can also catch invasive ants and even counterattack parasitoid wasps, according to Shinji Sugiura. This can be risky, though, since these insects can easily bite and sting back. Probably the fly-catching pugs started with more sedentary prey, and shifted from feeding on plant material such as flowers. It is thought that after feeding on flowers, some ancestral pugs became partial to the amino acids in pollen, along with the use of snapping as a purely defensive behaviour. It is even thought that these carnivorous caterpillars have somewhat taken over the role of predatory mantids in Hawaii where there are no endemic species. According to M.L. Henneman, one pug species, *E. monticolens*, provides an example of another route from vegetarianism to carnivory: it devours leaf galls and the tiny bugs living inside them.

Daniel Rubinoff also made some interesting discoveries in the Hawaiian archipelago when working on the massive radiation of *Hyposmocoma* moths (Cosmopterigidae; over 350 species are known!). These hyperdiverse moths show an amazing range of larval habits and shapes of larval cases, resembling candy wrappers, cigars and burritos. Some larvae are amphibious and can even live in brackish pools. In 2005, Rubinoff announced that larvae of four species attack and eat live snails. *Hyposmocoma pupumoehewa* (literally 'snail's nightmare') wraps a snail in a web of silk before breaking into its aperture and slaughtering it. Another species, with a similar gourmet taste for *escargots*, is aptly called *H. molluscivora*.

A TASTE FOR WAXY INSECTS

A number of caterpillars feed on scale insects, e.g. those of the cosmet moth *Coccidiphila gerasimovi* (Cosmopterigidae), the blastobasid *Holcocera iceryaella* and the boletobiine erebid *Eublemma scitula*, or exploit their secretions. *Eublemma amabilis* is considered a pest, predating on the scale *Kerria lacca*, an important source of resin for shellac in India. Carnivores of scale insects also include caterpillars of the

ABOVE One of only three extant specimens of *Euclemensia woodiella* collected by Robert Cribb in Manchester in 1829, a species not found elsewhere since. The moth's close relatives occur in North America, where their larvae are carnivorous on scale insects (*Kermes*) which contain a red dye and live on oak bark.

family Cyclotornidae from Australia and the North American cosmopterigid genus *Euclemensia*, at least one species of which is carnivorous on *Kermes* scale insects, which are used in the dying industry. *Euclemensia* includes the famous Manchester Moth (*E. woodiella*) of which only three specimens are preserved. The Texas moth (*E. schwarziella*) is a close relative if not conspecific. It may be that the moths were imported among a colony of scale insects by the cotton dying industry. Apparently, it was an 'intemperate' amateur moth collector Robert Cribb who, in 1829, found a large number of Manchester Moths on Kersall Moor flying around a hollow ash tree. The appearance of as many as 50 originally as the story goes (most of which were reputedly thrown in the fire by his landlady when he defaulted on his rent), requires a good explanation as to how they got there. It may not therefore be a coincidence that Cribb was a textile worker in the cotton industry. *Kermes* scale insects produce a dull red dye and maybe living *Kermes* were sometimes imported to Manchester from a port like New Orleans. The usual explanation proposed, however, is that the moths were imported in oak bark used in the tanning industry, maybe sometimes along with *Kermes* or *Allokermes* scales.

It is some unusual moth caterpillars indeed that graze (if not suck) the waxy bodies of true bugs. One of the more bizarre and rarely seen moth families feeding on Homoptera is the Epipyropidae, containing about a dozen genera. Some epipyropid species like *Agamopsyche threnodes* from Australia are parthenogenetic, i.e. they have only females, which generate female offspring. These epipyropids are rather small, inconspicuous, dark grey moths often with comb-like antennae, highly reminiscent of the adult males of some bagworms. *Ommatissopyrops lusitanicus* occurs in the Mediterranean, feeding exclusively on the Dubas Bug (*Ommatissus binotatus;* Tropiduchidae), which in turn feeds on palm trees. Other epipyropids feed as larvae on lanternflies, cicadas and their kin. *Epiricania hagoromo* feeds also on Ricaniidae bugs as well as on Cixidae and Dictyopharidae. Epipyropids are not usually carnivorous, but rather feed on the white, filamentous waxy secretions that occur on many of these bugs.

Some epipyropids, while generally quite modest in size, produce copious amounts of eggs (*Fulgoraecia* may produce 3,000), improving the chances that their hatchlings encounter a host, because the eggs are not laid directly on it but on the hostplant, which, however, will not be used as a foodplant. Epipyropids share their hostplants with their plantsucking bug hosts, also laying eggs on the foliage, which explains why they have particular hostplants at all. The tapered first instar is leech-like with powerful thoracic legs and after emergence, the young epipyropid caterpillar rests on the leaf edge waiting for its first victim to pass by. When the young larvae find a suitable host they crawl onto it, then metamorphose into a bizarrely corrugated form exuding a thick wax layer. Their needle-like mandibles project from and retract into a fat round head. It was often thought that epipyropids are not true parasites and relatively harmless to the host bugs, as they may just graze on their wax. Indeed, some hosts are relatively unaffected, but a number of epipyropids start grazing on the secretions and end up killing the host. Some species also feed on hosts that have no wax. In 1876, *Epipyrops anomala* was recorded feeding on the cottony secretions of the lanternfly *Pyrops candelaria*. Likely, it later fed on this bug's haemolymph using its sharp mandibles to penetrate the body wall. *Epipyrops eurybrachydis*, studied by B. Krishnamurti in 1933, definitely feeds on body fluids. Epipyropid larvae are often hard to spot on their Homoptera hosts, but when a larva is large a cue may often be an unusually raised wing of the host, under which the larva eventually shows up as a large waxy white lump.

CARNIVOROUS PLANTS AS FOOD

Carnivorous plants supplement their supply of nutrients, particularly nitrogen, by trapping a range of arthropods. Sometimes they ensnare moths. Perhaps the most famous example of species that turn the tables to feed on carnivorous plants is plume moths of the genus *Buckleria* (Pterophoridae), which specialize on sundews (*Drosera*). The biology of the Sundew Plume (*Buckleria paludum*) was discovered in 1906 by Thomas A. Chapman. Such sundew plumes must overcome the plant's glandular hairs whose sticky tips are death traps to many insects. Females lay eggs gingerly on non-glandular parts. The hatched larvae clear the hairs from a patch of leaf before making a meal of the leaf surface. In the process, they ingest the sticky droplets and not only preen themselves like a cat of any gum that adheres, but even eat some of the sundew's dead insects! Like sundews, they also secrete droplets at the end of their own specialized hairs. Other moths whose larvae feed on carnivorous plants are *Eublemma radda* (Erebidae: Boletobiinae), a scavenger of insects trapped by *Nepenthes* pitcher plants, and three species of *Exyra* (Noctuidae: Plusiinae), that eat *Sarracenia* flowers and trumpets. *Exyra* adults have specially modified pretarsal claws that enable them to run the gambit of the pitcher's

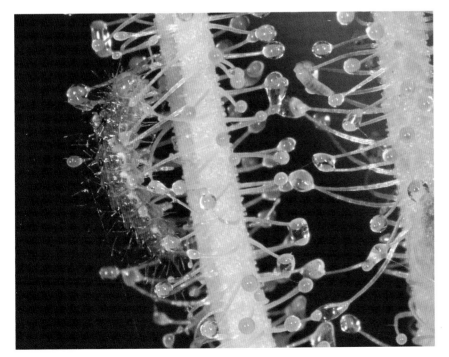

LEFT Larva of the Sundew Plume (*Buckleria paludum*) feeding on leaves of sundew (*Drosera filiformis*), a carnivorous plant. The caterpillar manages to avoid being ensnared by the plant's sticky droplets and even mops up this gummy substance.

treacherous internal walls, while caterpillars have an unusual series of projecting side lappets that prevent them falling down too far. As a last resort they can abseil back up to safety on a silk lifeline.

PIGGYBACKING ON SLOTHS

Apart from epipyropid moths and the earlier larval instars of cyclotornids (p. 71), the only other moths known to live directly on other animals are among the family Pyralidae. Pyralid moths of the subfamily Chrysauginae include the sloth moths, which 'piggyback' on sloths. The great naturalist Henry Walter Bates found the chrysaugine *Bradypodicola hahneli* in the fur of the Pale-throated Three-toed Sloth, (*Bradypus tridactylus*). Other sloth moths are *Cryptoses choloepi* and *Bradypophila garbei*, which live on the Two-toed Sloth (*Choloepus didactylus*), and probably also the Brown Three-toed Sloth (*Bradypus variegatus*). The sloth moths find the algae-encrusted hairs of sloths, which make sloths look greenish and camouflaged, an ideal environment to mate in, but they do not actually feed on the algae, as once suspected. In fact, it is not known what the sloth moth uses its short proboscis to feed on. A single sloth can easily carry a cargo of 100 moths, which are very active and probably use their peculiar arrow-shaped heads to run freely in and out of the fur. Their full biology was finally unravelled by Jeff Waage. As the sloth comes down from the treetops to perform its usual weekly 'ablutions' on the ground, the moths on its fur excitely jump off and lay their eggs in the sloth dung, a delicacy for their larvae! True dung-feeding in moths, shown also by some tineid caterpillars which live on bat guano and larvae of the Large Tabby (*Aglossa pinguinalis*), is extremely unusual, compared to many adult butterflies which avidly feed on dung. However, a piece of the puzzle in the sloth moth story was

RIGHT Sloth moths (*Cryptoses choloepi*) scurry around in the fur of a Brown Three-toed Sloth. The mated females hang about until the sloth descends to the ground to go to its latrine, then the sloths carefully covers the dung with leaves. This species is a rare example of a caterpillar that feeds on dung, but once the moths finally emerge, they must fly up to locate a sloth in the rainforest canopy.

missing. Why on earth would the laggard sloths face the risk of climbing down the trees to go to the loo? This exposes them to predators, so why do they not poop from the treetops? An explanation was found by J.N. Pauli and collaborators, who discovered that three-toed sloths in particular thrive if they enrich their diet by licking their fur algae, which is very rich in nutrients, carbohydrates and lipids. In turn, the algae benefit from the dead bodies of moths as a source of nitrogen. The sloths thus assist the life cycle of their guests by providing a latrine on the ground and are rewarded with fur infestation including enhanced algal growth. Another related moth from Brazil is even more extraordinary in its piggyback lifestyle. In 1926, Karl Jordan recorded larvae of the chrysaugine moth *Sthenauge parasiticus* feeding as ectoparasites (external parasites) on the branched tubercles of living larvae of the saturniid moth genera *Automeris* and *Dirphia*. A moth caterpillar feeding on a moth caterpillar!

CLOBBERING CLOTHES

Moths have a bad reputation as beastly insects that scuttle away when we open our wardrobe. With a sinking feeling we realize there will be clothes and other woollens shot through with holes. Even the most ardent moth apologists cannot force you to sympathize with clothes moths, but we can reassure you that they actually do a very important job. In fact, life on Earth only keeps on going because the waste substances of living organisms and the corpses of dead ones are broken down into simpler and simpler compounds. These are then available for assimilation again into the bodies of other organisms, thus renewing the living biomass. This essential 'composting' job is performed by organisms that are collectively known as decomposers. Some substances are fairly easy to decompose. For instance, you would hardly see remnants of a cow pat surviving for more than a year in a pasture. But there are other organic materials that are far harder to demolish, and may have relatively few organisms ready to banquet on them. One of these substances is keratin, the protein (actually a set of similar proteins) lining skin of tetrapod vertebrates and making up hair, fur, feathers, epidermal scales, hooves, claws and the outer part of horns. There are really few organisms capable of digesting keratin and a leading role in this task is taken by caterpillars of Tineidae, a most amazing family of micromoths in terms of diversification of diets, not just clothes. Tineids are therefore busy in nature removing keratin derivatives from carrion, nests, mammal holes and raptor pellets. Birds' nests are a great place to look for the diverse relatives of clothes moths. So much so that artificial birds' nests have been used in the tropics to lure tineid moths, an invention of the late Gaden Robinson, whose father designed the famous Robinson moth trap.

Some tineids have specialized on feeding on horns, such as the Horn Moth (*Ceratophaga vastella*), whose larvae feed, in Africa, on the hooves and horns of dead ungulates (not a natural group of mammals but including antelopes and their hooved ilk). A close relative, *Ceratophaga vicinella*, has been found feeding on keratin plates of shells of dead Gopher tortoises in Florida. If these moths tend to live in wild, open habitats, why do we find them hiding in the darkness of our clothes closets? Simply because we use wool so much, a fabric they are particularly

ABOVE Pupal cases of Horn Moth (*Ceratophaga vastella*) festoon Water Buffalo horns in the Kruger National Park, South Africa. Caterpillars of this species have special enzymes to digest horn keratin very effectively and thus they play an important role in nutrient recycling in the savannah ecosystems.

RIGHT A greatly magnified Clothes Moth (*Tineola pellionella*) among wool. Once an infestation starts, these moths are not easy to get rid of.

partial to, which is almost pure keratin, and also because naturally they would hide in dark places like inside bird nests. A few keratin-feeding species such as the Tapestry Moth (*Trichophaga tapetzella*), took the opportunity of joining us in our homes and have spread with us all over the world. Since the introduction of synthetic fabrics, *T. tapetzella* has become one of the rarest clothes moths in Britain. Interestingly, this species has some close relatives that keep on doing the same job in the wild (some of them specializing in owl pellets instead of making the leap into becoming synanthropic – that is, living in association with humans). Another two widespread clothes moths, *Tinea pellionella* and *Tineola bisselliella*, have a broader diet and can feed on other materials (the latter on silk) and therefore had added incentives to share our homes. How can we get rid of them? Only with difficulty. You can apply lavender spray, a lot of hoovering in hidden crannies, replacing carpets and clothes with synthetic ones, deep freezing and storing any potentially infested clothes in large sealed bags. Alternatively, obtain packs of sticky paper loaded with pheromone (see Chp. 4) placed in dark places near the floor so that females can also get stuck to them, are a few short-term remedies.

FUNGUS MOTHS

Fungivores are rare among moths. It is thought that among the most primitive living moths from the family Micropterigidae, some genera which have soil-living larvae may feed on fungi. The most common known examples of fungivores are again in the clothes moth family (Tineidae). A good example is *Morophaga choragella* in the subfamily Scardiinae, which is the major group of moth fungivores. Another group of tineids specializing on fungi is the subfamily Nemapogoninae; these moths can often be seen in woodland fluttering around fungus-infected trees. Their most well-known member is perhaps the Cork Moth (*Nemapogon cloacella*) that prefers the Oak Mazegill Fungus (*Daedalea quercina*) and Birch Polypore (*Fomitopsis betulina*). However, when these fungi are not available, the larvae behave like those of the Cork Moth's relative the Corn Moth (*Nemapogon granella*), a pest of dried store products, feeding on a wide range of other foodstuffs. The Wine Moth (*Oinophila v-flavum*) likes to feed on mould in damp cellars and sometimes drills into mouldy wine bottle corks, causing potentially expensive losses. Another tineid moth, *Amydria anceps,* is unusual in that its caterpillars actually eat the fungus cultivated by leaf-cutter ants, *Atta mexicana.* Heath Knot-Horn (*Apomyelois bistriatella;* Pyralidae) feeds on the polypore fungus *Hypoxylon occidentale.* The Waved Black (*Parascotia fuliginaria;* Erebidae: Boletobiinae), similarly feeds as larva on the underside of Birch Polypore, the bracket fungus once used for making the strips loved by microlepidopterists for double mounting of small moths.

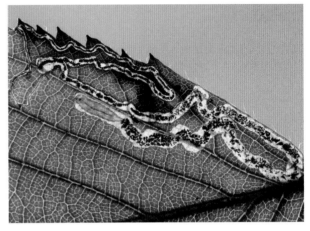

LEAFY LARDERS

It has recently been discovered that some leaf miner caterpillars such as those of the Spotted Tentiform Leafminer (*Phyllonorycter blancardella;* Gracillariidae) exploit their internal bacterial infections to prolong their leaf salad late into the autumn. These bacteria, in the genus *Wolbachia,* live inside the cells of their hosts and are transmitted from mothers to offspring via eggs. They have pervasive and remarkable effects on both larvae and adult moths in general. They can skew sex ratios, affect the variation of local populations, and now they are known to keep caterpillar food fresh. It has long been observed that leaf mines, even on abscized leaves scattered on the ground, have green islands, sections of leaf that stay green even after the leaves begin to senesce, so extending the period during which leaf miner caterpillars can feed. By giving female moths antibiotic-laden sugar water, researchers found that the disinfected lineages were no longer able to keep sections of the leaf mine green. *Wolbachia* interfere with plant metabolism in the leaf (specifically the plant hormone cytokinin) and green larders allow caterpillars sometimes to reach a third generation.

TOP The caterpillar of an oecophorid moth, the Golden-brown Tubic (*Crassa unitella*) tucks into the gills of a (mildly hallucinogenic!) fungus (*Panaeolus fimicola*).

ABOVE The tiny, flattened larva of the Golden Pigmy (*Stigmella aurella*) spends its life within a bramble leaf (*Rubus fruticosus*), leaving a tell-tale central 'frass' trail as it grows. This is one of the most familiar and distinctive leaf mines in Europe.

SELF-MEDICATING CATERPILLARS

Some moths seek special compounds known as pyrrolizidine alkaloids (PAs), which they normally get from plants such as the dying leaves of heliotropes (Boraginaceae). These include members of the tiger moth genus *Amerila.* As is known for sick monarch butterflies, some tiger moth caterpillars are thought to self-medicate. Michael S. Singer recognized these as cases of 'zoopharmacognosy', in analogy to

ABOVE Larva of an arctiine moth (possibly *Nyctemera* sp., or related genus) from Tanzania feeding on flowerheads of a tassel flower, *Emilia* (Asteraceae), a plant rich in pyrrolizidine alkaloids, especially in the flowers. Such moths can self-medicate on such plants.

other animals which use medicinal plants. *Apantesis incorrupta* caterpillars infested with parasitoid tachinid flies have been seen snacking on PA-containing plants like rattlepods (*Crotalaria*) and ragworts (*Senecio*). The caterpillars tolerate these plants as a 'bitter pill' instead of their normal hostplants (which are much more nutritious!), as the PAs give the growing parasitoid wasp grubs a nasty infusion that reduces their fitness, whilst boosting the caterpillars' immune system. Furthermore, it has been found that caterpillars of the Gypsy Moth (*Lymantria dispar*) can suppress transmission of highly contagious viruses when they eat leaves with high concentrations of toxins. Search and assimilation of particular chemicals other than as food to boost specific functions has been termed 'pharmacophagy', a phenomenon investigated by Michael Boppré. His studies on *Creatonotos* tiger moths demonstrated that male caterpillars feeding on PA-richer plants boost their adult sexual prowess allowing them to pump out remarkably large and alluring tubular sacks from the abdomen (these are called coremata, see p. 83) at dawn.

RAIDING THE NESTS OF WASPS, BEES AND ANTS

In their relationships with Hymenoptera moths display an exceptional spectrum of some of the most fascinating larval feeding relationships and coexistence with their hosts. Commensalism describes a loose association in biology where one partner benefits and the other is not actually harmed. Many moth associations are commensal but verge on parasitism, where the host can actually be harmed. A familiar 'wasp moth' is the Bumblebee Wax Moth (*Aphomia sociella*; Pyralidae). Caterpillars of this species, whose adults can play possum when disturbed, live in a wide range of nests of Hymenoptera. They not only clean up nest debris but may help themselves to eggs and grubs. The hosts sometimes generously extend their nests to accommodate the caterpillars, which protect themselves against stings by

RIGHT The Death's-head Hawk-moth (*Acherontia atropos*) is well protected against stings of bees when raiding their hives for honey, which is the sole food of adults.

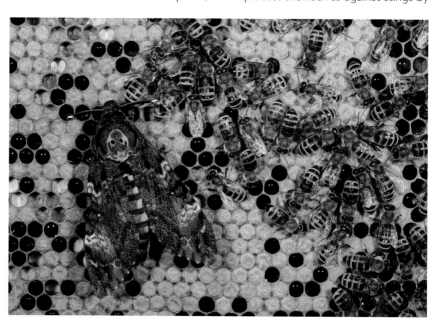

living in strong tubular cocoons of silk. In an intricate link with their nest environment it is thought that the females can synthesize their sex pheromone (see p. 79) using a chemical from an *Aspergillus* fungus found in honeycombs or in honey regurgitated to them by their wasp or bee hosts.

There are a range of moth families, including Psychidae, Tineidae, Cyclotornidae, Cosmopterigidae, Pyralidae, Erebidae (Arctiinae) and Noctuidae, which are associated with ants. These are known as myrmecophiles ('ant lovers'). Their relationship may cause some harm to the ants, e.g. the caterpillars eating the ants' brood. Naomi Pierce documented 130 moth species that are dependent on ants, either as obligate predators or parasites. Compared with commensals, infrequent visitors are known as myrmecoxenes ('ant guests'). Ants are inherently hostile to intruders, but frequently myrmecophile caterpillars (notably those of blue butterflies; few moth examples are known, see p. 68) secrete appeasing substances from special glands that reward the ants for not attacking them. Such ants regularly milk the caterpillars and in return even defend them against enemies.

There are a number of examples of loose associations with ants. This may sometimes just allow a caterpillar to feed with impunity rather than on an unusual food source. Living on the same tree as vicious weaver ants (*Oecophylla* spp.) in Southeast Asia and Australia (but capable of carrying on with their life perfectly well even when there are no ants), are the larvae of *Homodes* species (Erebidae). These caterpillars are strictly herbivores and even if they feed on leaves wrapping *Oecophylla* nests, the ants ignore them, as one of the authors (DCL) once noticed in Kuala Lumpur for *H. bracteigutta*. The caterpillars bear bulbous-ended filaments at either end of the body, which may either be glandular or tactile in nature, but are jerked incessantly. The caterpillars seem to be treated as kin.

Some moths actually live in ants' nests, in one or more life stages, where they simply feed on refuse. Examples are found in the tineid genera *Myrmecozela* and *Atticonviva* and a genus of furry-booted snout moths, *Pachypodistes* (Pyralidae: Chrysauginae).

Other such nest-scavengers are found in the bagworm genera *Iphierga* and *Ardiosteres* and the erebid genus *Idia*. Another such scavenger was reported in 1955, by Rector Renaud Paulian, who reared the giant Madagascan 'micromoth', *Acracona pratti* (Pyralidae) from the nests of *Crematogaster* ants. Even in Europe, it is not widely appreciated that the Dotted Chestnut (*Conistra rubiginea*; Noctuidae) oviposits on the trunks of deciduous trees where the ant *Lasius fuliginosus* is present. Although they can complete their growth on leaves, most end up in ant nests where they feed on insect debris such as dead ants and pupate there. Another family, with extraordinarily long tails, which

BELOW The weird larva of *Homodes bracteigutta* is strictly vegetarian, feeding on a range of trees. It is immune to attacks of vicious weaver ants (*Oecophylla smaragdina*), which, bizarrely, it resembles at both ends, enabling it to eat the ant-patrolled leaves with impunity.

ABOVE Larva of an ant-associated moth *Nudina artaxidia* on a trunk of Japanese Elm (*Zelkova serrata*) surrounded by *Lasius capitatus* ants. These larvae steal honeydew from scale insects that the ants also milk, while the caterpillars are completely ignored by the ants.

BELOW Larvae of *Eublemma albifascia* engage in mouth to mouth regurgitation (trophallaxis) from weaver ants (*Oecophylla longinoda*) whose nests they inhabit in West African rainforests. The caterpillars have extended true legs that enable a tactile exchange with the ants to encourage them to provide ant brood, which they eat.

might be associated with ants is Himantopteridae. Henry Elwes in 1891 reported finding a freshly emerged female of *Himantopterus dohertyi* in India that had just crawled out of an ant nest, its tails intact, while William Doherty had suggested to Charles Oberthür that *H. fuscinervis* in Java had an association with termite nests. This seems doubtful as the latter species has been reared by Henry Barlow entirely on the leaves of *Shorea platyclados* (*Dipterocarpaceae*). Across the tropics, species of *Acridotarsa* (Tineidae) scavenge in termite nests and some, in Africa, have peculiar lateral appendages and abdominal appeasement glands. More work on the feeding habits of loosely ant- and termite-associated caterpillars is clearly needed.

Some moths are certainly parasitic on ants. These obligate myrmecophiles directly depend on ants in some way. The pyralid genus *Niphopyralis* predates on ant broods. *Niphopyralis aurivillii* lives in the nests of *Polyrachis bicolor* in Java where, by mimicking the chemical signals of ants, its larvae get complete freedom to snack on the brood. Larvae of *N. myrmecophila* do the same in the nests of the Green Weaver Ant (*Oecophylla smaragdina*). The case-bearing larvae of the tineid genus *Hypophrictis* also feed on ant broods while one species, *H. dolichoderella*, just scavenges in bumblebee nests.

Some moths have more complex relationships with both ants and a third party. Takashi Komatsu and Takao Itino reported in 2014 that the footman moth *Nudina artaxidia* in Japan wholly depends on *Dendrolasius* ants but does not eat them. Rather, it is defended by the ants and is resistant to their toxins. The caterpillars even follow ant trails. They actually solicit release of honeydew, apparently their main food source, which is produced by scale insects that are attended by ants, and are themselves also free of ant attack. Belonging to the strange Australian moth family Cyclotornidae, the first instars of *Cyclotorna* prey on various families of bugs and are adopted by *Iridomyrmex* ants that enjoy their larval secretions. The late instars of *Cyclotorna monocentra* invade the nests of Meat Ant (*Iridomyrmex purpureus*). The legendary Australian lepidopterist or 'Kuranda butterfly man', Frederick Parkhurst Dodd, even noticed that cyclotornid females seem to cue in on ant trails and actually lay their eggs near neighbouring plant hoppers. After they hatch, the strange-looking bug-eating larvae which themselves resemble scale insects wait for an ant to pass. When ready, they curl up to expose an appeasing anal secretion and this solicits the ant to carry them to the nest.

An association between moths and ants has developed in one case into a cuckoo lifestyle. The larva of the marbled moth (*Eublemma albifascia*), which was studied by Alain Dejean, lives in the nests of the West African weaver ant (*Oecophylla longinoda*). The moth eggs are directly laid on the silked-together leaves that provide the building framework for the ant nests and the eggs are quickly transported by the workers, which place them among the brood. The later instar larvae are very unusual in that they have elongated forelegs and trigger mouth to mouth regurgitation (trophallaxis) from weaver ants within their nests, even stimulating them to supply ant eggs. It is thought that the ability to feed on coccids and other bugs was a prerequisite to the evolution of this extraordinary lifestyle. Somehow the caterpillars acquire the colony odour and can feed with impunity. Such cuckoo-like behaviour was only previously reported in lycaenid butterflies. One colony of weaver ants had 359 caterpillars in residence! The term 'cuckoo' is appropriate because caterpillars in such numbers can actually 'ruin' a colony. Not only is the sex ratio of ants produced being altered but the queen is being neglected and in some cases starved to death!

One South African cosmet moth (*Coccidiphila stegodyphobius*; Cosmopterigidae), is thought to be actually commensal with a *Stegodyphus* spider, feeding on dead scraps of animal matter, yet skilfully avoiding getting caught in the spider's web. Henri-Pierre Aberlenc also found another cosmet moth among the communal cocoons of processionary moths (*Hypsoides* sp.) in Madagascar. David Agassiz has found exceptionally diverse, partly commensal moth assemblages in ant-hosting *Acacia* thornbushes in Kenya.

ADULT FEEDING BEHAVIOUR

Though adult moths are not streamlined into pure 'eating machines' as are caterpillars, if they have not stocked up the necessary reserves to power the adult stage, the adults still need to feed or seek additives not in the larval diet, and subsequently have developed some interesting strategies. Some feed on spores and pollen, while most drink nectar, even from the flowers of the caterpillars' hostplant. A few sometimes drink tears and other exudates from vertebrates. Natural selection has led to some incredible adaptations, not only in mouthparts but also in lifestyles which we discuss below. Some of the most remarkable adult feeding adaptations occur in the yucca moths.

MOTHS THAT FEED ON SPORES AND POLLEN GRAINS

Moths in the family Micropterigidae have a remarkable cavity in the mouth that functions much like a bird's gizzard. Known as a triturating basket, it grinds up spores. Adults of the most primitive micropterigids, mostly in the Southern Hemisphere, gather under fern fronds or occasionally on flowers where they can find a mate and spend the rest of their time chomping on fern spores (or pollen). In their primitive, five segmented, folded condition the maxillary palps of micropterigids (we have already discussed on p. 20 how these are reduced in more advanced Lepidoptera), delicately guide the spores towards the mouth, the tip of the last segment having mushroom- or paddle-like bristles for spore attachment. Species of the South American family Heterobathmiidae, which are unique to Patagonia, feed on the pollen of their southern beech hostplants, *Nothofagus*. In a similar way to micropterigids, they use their maxillary palps to manipulate pollen and break it with their mandibles, so they can grind it in their triturating baskets.

Ever the messy eaters, adults of the yucca moths in the genus *Tegeticula* (Prodoxidae), end up actively collecting pollen as they feed and carry some to the next flower they visit. In this way they serve as specialist pollinators of yucca plants. They are perhaps the best-known case in moths of mutualistic 'symbiosis' (literally 'living together'), because the moths get food in return for their pollination service, with the plants 'paying' for the service with the loss of some seeds to the moths' larvae. Not all yucca moth examples though are truly reciprocal. Some species, the so-called 'bogus yucca moths' of the genus *Prodoxus*, 'cheat' on the plants as they do not assist pollination but only feed on them. John Thompson and Olle Pellmyr dedicated years of research to study the complex co-evolutionary relationships between prodoxid moths and their hostplants, and it appears that co-adaption and diversification have been driven by the interaction. *Greya* species, primitive prodoxids, range from loose to strong mutualistic association with their hostplants, which are almost entirely saxifrage plants in the genus *Lithophragma*. Prodoxids symbiotic with yuccas, notably *Tegeticula*, have some very special adaptations to feed on pollen. They produce enzymes (sporopolleninases) that break down the amino acids in pollen. The adaptations include also very striking morphological specializations. In female yucca moths, the base of each maxillary palp branches off into a long 'tentacle', which remarkably can also be coiled, as if the palp bears an additional proboscis. It is thought that such a bizarre adaptation acted as a 'key innovation' in the history of the mutualistic interaction between yucca moths and yucca plants. This specialized adaptation enables the moths to collect a huge pollen load, which

BELOW The Plain Gold (*Micropterix calthella*) is one of the more frequently encountered Northern European micropterigid moths, but being tiny its golden scales are rarely photographed at such resolution. The adults are very partial to the flowers of buttercups and Kingcup, in which they forage for pollen and often aggregate in the late spring.

LEFT A heliozelid micromoth feeding on pollen of *Boronia crenulata* in Western Australia, a plant with which the moth has a mutualistic pollination relationship. This is one of the most recent such discoveries.

can be easily seen in their mouthparts and which the female will carefully pack onto the floral stigmas.

The study of co-evolutionary relationships between moths and flowering plants is a rapidly developing field. Liz Milla, Doug Hilton and their co-workers recently discovered heliozelid moths (Heliozelidae) that are not only symbiotic but mutualistic in their relationships with *Boronia* flowers (Rutaceae) in Australia, assisting in their pollination. Females have remarkable structures on the underside of the abdomen, which enable them to gather pollen.

Another research programme into mutually beneficial relationships between micromoths and plants is that of Atsushi Kawakita and Makoto Kato who studied the genus *Epicephala*, a group of gracillariid moths strictly associated with plants of the family Phyllanthaceae. Female moths are attracted by night to the flowers and lay eggs on the stigma, fertilizing the flowers with pollen from another

flower in the process. In some species, the larvae leave a certain number of seeds in the fruit to develop, but in other species, they are more harmful in the relationship, devouring the entire cache of seeds. Cases of such seed robbing *Epicephala* are rare because they inflict a heavy cost on the plants despite any advantage of pollination, a strategy which in the long run may not benefit the

ABOVE Female of *Epicephala perplexa* pollinating a flower of *Glochidion acuminatum*. In return for its pollination service, the plant sacrifices a proportion of its seeds to the moth larvae.

moths either. Nevertheless, such is the importance of *Epicephala* pollination to the Phyllanthaceae, that comparing the evolutionary history of these plants to that of moths, the researchers discovered that their mutual relationship has evolved at least five times independently.

NECTAR FEEDING ON ORCHIDS

The hawkmoth known as Wallace's Sphinx (*Xanthopan morganii praedicta*) is nectar drinking and is known to visit the spectacular flowers of Darwin's Orchid (*Angraecum sesquipedale*) in the Madagascan rainforest.

These flowers are extraordinary. In 1862, in his treatise on the pollination of orchids, Charles Darwin himself thought that a hawkmoth might be responsible for the pollination of the 30 cm (12 in) long flowers and famously proclaimed 'Good Heavens what insect can suck this?' In 1867, Alfred Russel Wallace stated memorably "One from tropical Africa (*Macrosila morganii*) [the hawkmoth's name at that time] is seven inches and a half. A species having a proboscis two or three inches longer could reach the nectar in the largest flowers of *Angraecum sesquipedale*, whose nectar spurs vary in length from ten to fourteen inches. That such a moth exists in Madagascar may be safely predicted; and naturalists who visit that island should search for it with as much confidence as Astronomers searched for the planet Neptune, - and I venture to predict they will be equally successful!" It was not until 1903 though that Karl Jordan and Lionel Walter Rothschild actually described a population of that moth from Madagascar. Thus, as dramatically as his quote, Wallace had predicted the existence of this exact species of moth 36 years before its discovery to science, as originally envisaged by Darwin.

In 1992, a pollen mass of the orchid was found adhering to the moth's proboscis for the first time by Lutz Wasserthal. In 1997, he managed to photograph the pollination event by caging out moths and flowers in Madagascar. Then in 2004, after staking out a flower of *A. sesquipedale* for three nights in a row, Philip DeVries finally was filmed observing a Wallace's sphinx visiting an orchid in the wild, thus proving the theory that had been proposed by Wallace 137 years earlier. The moth's proboscis is 14.5–24 cm (5.7–9½ in) long. However, it has been known for a long time that there is another hawkmoth in Madagascar, *Coelonia solani*, which exhibits a similar range of proboscis lengths. Is it a competitor and the result of a race for the longest tongue that can suck the last drop of this orchid's nectar? Apparently not! Actually *C. solani* seems to avoid *A. sesquipedale* and instead, offered the choice, likes to feed from the most closely related orchid, *A. sororium*. Along with other plants, notably in the families Rubiaceae, Caprifoliaceae and Solanaceae, *Angraecum* orchids are thus another group of 'sphingophilous' (sphingid-pollinated) plants. However, in the wild the story is doubtless more complicated, and the two Madagascan hawkmoths with giant proboscides may indeed compete. Furthermore these moths are not limited to feeding on orchids. It has been discovered by botanist David Baum that both species are the most important pollinators of *Adansonia perrieri*, a baobab species found in western Madagascar.

These hawkmoths have a fascinating behaviour. They have a technique called swing-hovering. Swinging from one side to another, they hover at an orchid with the tip of their proboscis inserted into the flower spur, allowing them to be wary of predators. Remarkably, lemurs have learned to snatch the moths when these are nectaring on baobab flowers. When servicing the long-tubed *Angraecum* orchids, the proboscis is usually long enough to get at the nectar right at the bottom of the tube, and allow the plant to glue the pollen package on to the base of the proboscis ready to be transferred to the stigma. *Angraecum* species with longer nectar tubes are increasingly reliant on such a powerful disperser as *Xanthopan morganii praedicta*, and avoid less dependable pollinators. This can work because the longer the tube, the larger winged and more powerful a flyer the moth must be, due to a strong biometric correlation between wing size and proboscis length. Sometimes the spur of an orchid can be considerably longer than the known range of proboscis lengths of any known giant hawkmoth in Madagascar, as in *A. longicalcar*, which has a 30–40 cm long nectar tube. This has encouraged

LEFT Wallace's Sphinx (*Xanthopan morganii praedicta*) is the sole pollinator of Darwin's Orchid (*Angraecum sesquipedale*) in Madagascar. The moth has a proboscis nearly the length of the huge spur of the orchid, enabling it to reach the last drop of nectar whilst the pollen masses are being attached.

speculation among zoologists about a now extinct pollinator! Perplexingly, camera traps placed by botanists around the tiny remaining population of this critically endangered plant have failed to find any visitations. Here is an example of an exquisite orchid in an evolutionary dead end in Madagascar for which we have not even observed its moth pollinator, if it still survives.

DRINKING TEARS OF SLEEPING ANIMALS

Madagascar is the setting for another really remarkable moth story. The world's first bird-'eating' (drinking) moth was discovered in February 2004 by a team of ornithologists (led by Roland Hilgartner) who were studying sleeping birds in the forest of Kirindy. Large moths were photographed clearly feeding on both the Madagascar Magpie Robin and, in December 2005, on the Common Newtonia. The moths fed at night for up to 30 minutes, without either bird batting an eyelid. With closer observation it was seen that the moth inserted its proboscis under the closed lids of the bird's eye. Hilgartner managed to catch a specimen and send it to one of the authors (DCL) for identification. The moth proved to be a male of *Hemiceratoides hieroglyphica* (Erebidae) and close examination of the proboscis revealed the tip to be forked and have vicious backward-pointing barbs and spines that latch between the membranes of the bird's eye, apparently to suck salts from the tears.

This was the first known observation of moth attraction to eyes (opthalmotropy) in Madagascar, where hoofed mammals are naturally absent. Ungulates are the usual hosts of tear-drinking (lacryphagous) moths, of which many examples are known around the globe, notably in the families Geometridae, Erebidae, Notodontidae,

RIGHT Harpoon Moth (*Hemiceratoides hieroglyphica*) feeding on tears of a sleeping Common Newtonia in Madagascar. This photograph represented the first observation of a 'bird-eating' moth, but the phenomenon of moths feeding on birds' tears, previously well-known on mammals, is becoming more widely noticed around the tropics.

Crambidae and Drepanidae. Some species have also been observed frequenting eyes of other mammals such as elephants and, not least, man. In Africa, such eye-frequenting moths repeatedly visit and may sometimes even benefit by digesting leucocytes from pus that presumably results from infection by bacteria or viruses! Hoofed animals are relatively defenceless against these nocturnal visitations, whereas primates can easily brush off irritating insects. Perhaps birds, which are also fairly defenceless, may prove to be more commonly attacked than previously thought. Indeed another such attack was filmed in Brazil of the erebid *Gorgone macarea* on a Black-Chinned Antbird. There is another *Hemiceratoides*, currently classified as a subspecies of *H. hieroglyphica* (*vadoni*), which is known from eastern Madagascar, and others in Africa, so tear feeding observations are needed in these moths, too. Finally, there is a potential vampire moth species in Madagascar, the erebid *Calyptra triobliqua*, which also has a viciously barbed proboscis. But what it feeds on is currently unknown! There is much still to discover in a place as strange as Madagascar, and elsewhere!

VAMPIRE MOTHS

The discovery in 1980 by Häns Banziger of the first known case of vampirism in a moth, *Calyptra eustrigata*, was astonishing. Some 18 species of *Calyptra* are now known and include about 10 capable of piercing the skin of mammals to feed on blood. Some moths will even feed on humans, including the Asian species *C. bicolor*, *C. fasciata* and *C. ophideroides*. There is one vampire moth species that occurs in Europe, *C. thalictri*, though somewhat reassuringly for Europeans, it has only been seen blood-feeding under experimental conditions by Jennifer Zaspel and Vladimir Kononenko in the Russian Far East. Hans Bänziger recounts how he was driving along in a taxi in Thailand and a moth flew in the open window and attacked his chauffeur. In order to feed on blood, a *Calyptra* moth moves its proboscis back and forth and drills it into sometimes thick skin. It then more emphatically rocks its head to ram it in further, before hooks on the moth's proboscis latch it in position. Then it sucks the blood, although it is not known if any anticoagulants are used. The evolutionary sequence is thought to be from piercing fruit to animal skin as the moths developed a taste for blood.

Many tropical moths do in fact pierce fruits to suck their juices. With strong tearing hooks on the proboscis, many beautiful large moths in the erebid genera *Eudocima* and *Phyllodes* (as well as *Calyptra*) can easily penetrate through the thick skin of fruits such as orange, pineapple or pomegranate. Some of these moths turn out to be major pests in orchards, as the wounds they make in the skins leave the fruits exposed to fungal and bacterial infections. There are, however, several less specialized moths that feed on fruits whenever they are given the opportunity. They usually visit soft-skinned or fully ripened, wounded fruits but a few species can pierce skins as hard as those of rose hips. One freezing Christmas night in the snowy environment of the Apennine Mountains, AZ was quite amazed to witness aggregations of *Conistra* moths frantically piercing the berries of a rose bush.

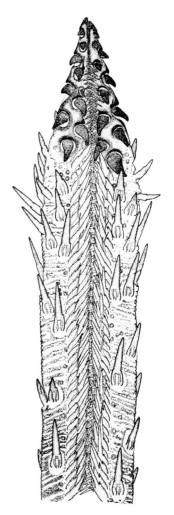

ABOVE Proboscis tip of a vampire moth (*Calyptra eustrigata*) showing tearing hooks at the top and erectile barbs.

GOING TO MORE UNUSUAL LENGTHS

Feeding on tears of sleeping birds or the blood of mammals is astonishing enough. However, since the dawn of their evolution, adult moths have been incessantly exploring new food sources incessantly and much remains to be discovered. In 1976 North American lepidopterist Mogens C. Nielsen found a hawkmoth, *Sphinx luscitiosa*, drinking the liquids of a decaying fish with its proboscis deeply inserted through the fish's skin. He also collated unusual records of other hawkmoths that had occasionally been spotted banqueting on fish, carcasses or dung. Jeremy D. Holloway recalls that in Borneo, geometrids of the genus *Zythos* have been attracted with carrion-baited traps. Quite long is also the list of species which have been seen drinking at animal wounds, exudates or even human sweat.

AZ once observed during daytime a swarm of normally nocturnal Light Brocade moths (*Lacanobia w-latinum*) around a black poplar tree. The tree was infested with Poplar Spiral Gall aphids (*Pemphigus spyrothecae*) which make the leaf stalks (petioles) swell and coil to form tightly spiralled galls protecting the bugs. Remarkably, the moths were frantically inserting their proboscides into the thin passageways left at the centre of these whorls to drink the bugs' honeydew.

We have seen the extreme diversity of resources in the moth larder. Now we move to the main goal of all this gluttony: to find the ideal moth mate.

RIGHT Nessus Sphinx Moth (*Amphion floridensis*) feeding on carrion. Dung feeding in adult Lepidoptera is more often noticed in butterflies. It is likely that, as in butterflies, the males crave a salt lick or seek nitrogen-laden nutrients.

ABOVE *Clelea formosana* feeding on freshly deposited dung in China.

LEFT A Costa Rican erebid (*Euclystis proba*) sips directly on the 'honeydew' secretions of a lantern bug (*Enchophora sanguinea*). Moths are attracted to honeydew from a range of bugs, including aphids.

CHAPTER 4

Mating

MOTHS USE ONLY two life stages (sometimes just the larval one) to feed, efficiently powering their development towards their ultimate biological goal, which is reproduction. Successful moth sex involves a combination of many different cues: olfactory, visual, auditory and tactile. There are four main phases: moths must meet; indulge in a courtship; mate (copulate) and then ensure successful fertilization. Furthermore, sexes may have to successfully spar among themselves. Some of the mechanisms involved can be quite elaborate. As we point out, love among moths is often quite different from the situation in butterflies where individuals mainly rely on sight.

ENCOUNTER STRATEGIES

Either female or male moths (or both) may take the initiative, most often using scent or seducing with an attractive visual display, sometimes also using sound. Since it is usually the females that will take the lead, males must do the travelling, sometimes from a considerable distance, and usually with some effort (flying into a headwind).

FEMALE INITIATIVE
In most cases, female moths seize the initiative in enticing potential partners, even from afar, because they have an irresistible perfume (the female sex pheromone) that they can waft to the wind. Pheromones are defined by analogy to hormones. Hormones elicit responses in target organs within the body. Pheromones are chemical signals – they are released by an organism and elicit responses in conspecific individuals – and in insects have a broad range of functions, not just sexual, as they can promote gregariousness, sociality, caste determination, sibling-recognition and alarm. The first pheromone ever to be characterized was bombykol, the sex pheromone of the Silkworm (*Bombyx mori*). It took 20 years' work by Adolf Butenandt and colleagues to assess the chemical structure of the compound. By 1959, using no less than half a million female abdomens, they got 6.4 mg of this precious substance.

OPPOSITE Vive la différence! A mating pair of tropical eggar moths (*Trabala* sp.) exhibits outstanding sexual dimorphism.

Assembling and matings of Silkworm moths (*Bombyx mori*). Some females (stouter-bodied) can be seen calling with their pheromone glands extruded.

Some of the most striking examples of pheromone-mediated long distance attraction are illustrated by virgin females of eggar and emperor moths, which 'call' at particular times of the day. Mere traces of their pheromones in a container can lure males from several kilometres downwind. Male moths just fly upwind and when they detect molecules of the female perfume on their antennae, they zigzag in flight as they lose and then retrace a whiff of the pheromone plume. Records of male moths that flew for 10 km (6¼ miles) or more are not uncommon. Jean Henri Fabre in the late 19th century described how a single female of the Giant Emperor Moth (*Saturnia pyri*) could attract 150 males over the course of a week. When he placed her under a bell jar, the suitors lost interest. This revealed that sight was not involved and was not the primary factor in attracting a mate. When Rudolf Mell in 1922 wanted to assess the distance of attraction in a moon moth (*Actias ningpoana*), he devised an ingenious experiment in China. He left virgin females in a gauze cage at his home and released marked males at intervals from a series of train stations, discovering that they could return from as far away as 11.6 km (7¼ miles).

BELOW Two colourful males of Emperor Moth (*Saturnia pavonia*) compete to mate with a virgin female (grey).

BELOW RIGHT Spicebush Silkmoth (*Callosamia promethea*). One ardent male of this species is reported to have flown 36.5 km (22¾ miles) in response to a whiff of pheromone. No doubt strong tailwinds play a role in assisting such flights.

Many decades of research have now been carried out on female moth pheromones and since the discovery of bombykol the pheromones of many species, especially pests, have been isolated and artificially synthesized in the laboratory. Some of these pheromones are now widely available and can be used to attract and monitor moths or disrupt matings of many species of agricultural and forestry importance.

The pheromone may be a single compound, sometimes made up of one of two possible mirror-image chemical structures (known as optical isomers), or a finely tuned blend. Such highly volatile substances are composed of a chain of between five and 23 carbon atoms. With the huge diversity of moths that fill the air, not every compound is specific. Some species share the same compound(s) in their female pheromones, but it is often the ratio of the mix that provides the right lure. Ensuring that males of the right species are attracted to avoid wasteful interactions often involves careful release at a particular time of day. This is why males of closely related species usually do not fly at the same time of day or night. Indeed, as expected due to the similarities, sometimes serious mistakes occur. They are rare, but there are many cases of matings not only between related species or genera, but completely the wrong family! Usually genitalia diverge enough between related species so as to make such matches impossible, but sometimes the male can effectively clasp the wrong female anyway. AZ has noticed in the Apennines that among all couplings of mimetic handmaidens (*Syntomis phegea*; Erebidae: Arctiinae) and burnets (*Zygaena* spp., Zygaenidae), which often share the same flower heads, up to 5% were with the wrong family. In this case it is likely that males get mixed up in close proximity, perhaps responding to the right pheromone but responding in a tactile fashion to the nearest partner. In 2012 though Yûsuke Kondo and colleagues provided experimental evidence though that the Japanese Nine-spotted Moth, (*Amata fortunei*; Arctiinae) can use visual cues as well as chemical ones to fine-tune finding the right mate among similarly coloured moths.

Sometimes other organisms can cleverly exploit female moth pheromones. There are actually spiders, bolas spiders (see p. 98), which have managed to mimic the pheromones of particular species of moths and lure males to their deaths.

ABOVE It is a rare phenomenon but coupling sometimes happens between different families of moths. A Six Spot Burnet (*Zygaena lonicerae*; Zygaenidae, top) mates with a Handmaiden (*Syntomis phegea*; Erebidae, bottom). Syntomine erebids can use visual as well as chemical cues, but this kind of mixup may be due just to physical contact when jostling on the same flower heads.

MALE INITIATIVE

The female calling pheromone has been lost in a few moths, but the males call loudly and shrilly enough that it is thought they can attract females from a distance. In Australia, males of the crambid moth *Syntonarcha iriastis* perch at the top of bushes

LEFT A male crambid moth (*Syntonarcha iriastis*) in Australia sings in shrill ultrasound resonating its impressive abdominal sacs, probably to attract females from a distance.

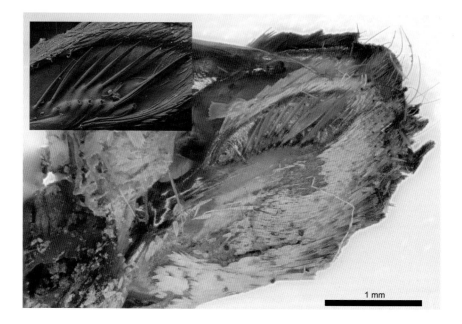

1 mm

and serenade using 42–57 kHz ultrasound. As in some hawkmoths which indulge in loud genital stridulation it appears that the genitalia bear the sound producing organs as a complex array of hardened flaps, file scrapers and resonators. Another crambid studied by Klaus-Gerhard Heller and Rüdiger Krahe, *Symmoracma minoralis*, behaves in Western Australia in a similar way, with a shrill chirping at 60 and 120 kHz. In this case, an expapdable structure at the tip of the abdomen broadcasts the sounds made by contractions of the genitalic tymbal organ of this species. These moths do not seem to form groups of perching males. Other species, by contrast, form a lek, which is a gathering of males that collectively signal their presence to females.

In some cases, moths may lek visually. The silvery white reflective males of the Ghost Swift (*Hepialus humuli*) appear at dusk, swaying to and fro in a phantasmagoric cloud, and females come to these leks. The males start their hovering bouts synchronously, usually as a handful but sometimes swarming in dozens. According to classic studies by Jim Mallet and John Turner, females are initially attracted visually, select a male from the lek and then engage them with a courtship ritual. Such a reversed calling system happens in many hepialids. Aerial knockdown by a female has even been witnessed. This interaction is mediated by another compound, a male sex pheromone this time, which is released from brushes on the hind-tibiae during flight. The male then pursues the female to a courting site on low vegetation. Other hepialid species without brushes just actively pursue the females in a less elaborate manner.

Males of the beautiful Green Long-horn (*Adela reaumurella*) are frequently observed to lek by swarming altogether in full sunshine on top of bushes and branches, and they can probably vie for position and spot visiting females with their near-holoptic eyes (see Chp. 1), likely an adaptation for swarming as they are in many Diptera. This behaviour occurs even in the most primitive moths. Males of the family Micropterigidae use either fern fronds or flowers as daytime hang-

outs for the meeting of the sexes. They also swing hover to display their metallic colours in flight, which may promote aggregation. DCL recently observed apparent lekking of metallic male micropterigid moths in the Anjanaharibe Sud forest reserve in Madagascar. From a globular swarm of three undescribed species, males of one species assembled under a fern frond bearing fertile spores (see p. 69); when females arrived they got a lot of attention from rivalling males.

Lekking may involve perfumes as an alternative to visual interaction or they may be combined with reflective scales. Micropterigids not only have glittery scales but also often have a specialized gland on the fifth abdominal segment in both sexes (a primitive characteristic shared with caddis flies), thought to facilitate their aggregation. The classic case of pheromones used in lekking is for the Indo-Australian tiger moth genus *Creatonotos*. Perching males expose collectively their hugely inflated coremata at dawn. These impressive tubular expansions of the abdomen release a scent. Females visit these leks and the males gain an advantage with a collective lure. The size of these organs depends on what the caterpillars eat, so the males manage to compete in this way (see Chp. 3, p. 66).

There seem to be few reports of moths lekking using sound. A classic example was published in 1995 by John Alcock and Winston T. Bailey in two agaristine noctuid moths, the Australian Whistling Moth (*Hecatesia exultans*) and the related *H. thyridion* (Noctuidae: Agaristinae). They found that males perch on bushes, in the manner of songbirds, 15–25 m (50–80 ft) apart, and call vigorously for hours with their forewing 'castanets'. This sound is ultrasonic (c. 30 kHz). They call for several weeks in a loosely defined area, hopping from perch to perch within a radius of some 10 m (33 ft) and defending these territories aggressively against any male interlopers, hoping that a female will choose them as opposed to other successful perch defenders. It is the female that solicits copulation, presumably choosing the male with the best 'physique' that wins enduring interactions with other males. After mating, she flies to other places in search of nectar and suitable hostplants.

BELOW Male of *Creatonotos gangis* extends its spectacular coremata at dawn in the Queensland rainforest.

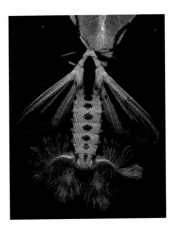

ABOVE A male of Whistling Moth (*Hecatesia exultans*) perches on a bush and makes a loud sound with its extraordinary castanet-like modification of the forewing margin. Interaction with other males results in fierce aerial duels. Within-species communication also involves chemical signalling, as the males are also equipped with sweet smelling scent brushes on the legs.

There are all sorts of sounds produced by males of different agaristines. These moths are also extremely energetic, and vigorously defend their patches. DCL noticed that the frenetic aerial chases of *Heraclia butleri* in the Kruger National Park, South Africa, are accompanied by a loud, rapid-fire clatter. Many sounds produced by agaristines remain unknown – no-one has ever reported the behaviour of *Pemphigostola synemonistis* from the isolated forests of southwestern Madagascar, but we can imagine it produces a loud noise considering the male's intricate stridulatory organ. The tibia of the midleg bears conspicuous transverse ridges which may be drawn, like the bow of a violin, across a transparent cupped structure large enough to distort the shape of the base of the forewing (see p. 6). Such a configuration was first noticed in this species (along with its apparent close relative *Musurgina laeta*), in 1921 by Karl Jordan. Males of sound producing Lesser Wax Moth (*Achroia grisella*) also sometimes aggregate into leks.

SPARRING MALES

In most cases such as the above, sparring in moths is confined to the male sex. Not only do some moths compete using loud sounds but with pungent scents. DCL noticed that male agaristine moths of the genus *Rothia* in Madagascar, which have extrusible thick scent brushes at the tips of their abdomens, scent-mark their perching

RIGHT *Urania leilus* male foreleg showing the position of stridulatory peg (Co = coxa, partially detached from the thorax; Fe = femur; Peg; Ti = tibia). In order to emit its clicks used to defend its perch, the male rapidly protrudes its forelegs brushing the coxal peg across the head of the femur.

Peg

territories with a sweet musky odour. These spots can be smelled by humans long after the moth has departed. Moreover, the musky aroma of these colourful moths can last in a dried specimen for many years. Male *Rothia* engage in frenetic aerial chases which are territorial in nature, usually after potential rivals, but presumably the goal of this behaviour is not just to defend a good patch in the forest, but to give chase to any receptive female that flies past.

One of the authors (DCL) studied the day-flying uraniid moths *Urania fulgens* and *U. leilus* in Panama and Peru. Holding the moths to his ear he noticed a fine buzzing noise. Sound production, while defending perches in male-male interactions of *U. boisduvalii* in Cuba, was also confirmed by Alejandro Barro. It turned out, after detailed examination by DCL, that males of all *Urania* species have a remarkable peg-like structure on the foreleg coxa composed of scales whose tips are hooked together to form a sort of plectrum that is rasped against the drum-like head of the femur of the same leg. A series of elastic deformations of this stoutly supported peg when scraped against the femur produces a train of audible clicks. The males while sitting head downwards on leaves were observed to make click-inducing thrusts of their forelegs in response to passing males, although the signal might also be perceived by passing females. The related Sunset Moth (*Chrysiridia rhipheus*) from Madagascar appears not to 'sing' and instead has some kind of a scent organ in a similar position to the *Urania* foreleg peg.

SPARRING FEMALES

Interestingly, it is not always males that compete with each other in the mating game. Females of the southern European species *Heterogynis penella* are wingless and legless, and after emergence, they remain physically bound to their pupal case. Accordingly, they must lure in the winged males. Due to their immobility, they are fixed to the exact position where they made their cocoon. So how do all these calling females compete for the wandering males especially when there is a crowd? They have to plan ahead and, as caterpillars, choose the right spot to make their cocoon. AZ noted that female caterpillars occurring in mixed scrubland-grassland in the Apennines and northeast of Italy compete by reaching elevated positions, from where as adults they can more easily spread their pheromones, ostentatiously exhibiting their vivid black-and-yellow striped pattern (probably also as a predator defence). As caterpillars about to spin their cocoons, they climb high up on grass stems. Male caterpillars do not bother to climb and prefer to hide their cocoons among low vegetation, as the resulting adults can fly.

BELOW The remarkable, yellow-striped, immobile 'appendage-less' larviform female of *Heterogynis penella*, freshly emerged out of its cocoon. The female can only successfully lure the males from the elevated site chosen to spin the cocoon when she is a larva.

SEXUAL SELECTION IN MOTHS

Visual cues play a major role in butterfly communication between conspecifics. Many of the colourful patterns seen in butterflies are considered to have been modelled by sexual selection, and butterflies in particular are frequently strikingly different between the sexes (sexually dimorphic, with the males more conspicuous). Sexual selection leads to the development of features that are advantageous in terms of achieving reproduction, sometimes at the expense of survival. Thus, there may be a trade-off between sexual and natural selection (both of course are 'natural' processes). A good example is the usually conspicuous shiny white colour of male Ghost Swifts. Why are they so conspicuous? This gives them a means to display easily to females at dusk, which initially select males visually (see p. 82). Interestingly, in the Faroe and Shetland islands, Ghost Swift males are themselves dimorphic, exhibiting also a brown variant. This brown form of the male better escapes predation from

seagulls and terns because it is much less visible at dusk, whereas the birds more intensely predate any white males.

There are two main types of sexual selection. Intersexual selection exists when an evolutionary trait is favoured because the bearer is preferred by members of the other sex (e.g. females choosing males with particular colour combinations). Intrasexual selection occurs when a trait allows the bearer to outcompete members of its own sex in getting access to mates. Males, the more dispensable sex, can 'fight' with 'show-off' features that better drive away competing males. These can be colours, scents, sounds, morphological, physiological and behavioural features. The two types of sexual selection are not mutually exclusive and sometimes, as before, there has to be some trade-off. For example, the size of *Creatonotos* male coremata may be driven in different directions by these two types of sexual selection.

COURTSHIP RITUALS AND SAFELY ENGAGING WITH THE OPPOSITE SEX

Once a male has found a mated female butterfly, such a female may need to take steps like raising her abdomen or using anti-aphrodisiac pheromones to avoid harassment. This is because visual cues may lead males to the female even if she has already mated. Most moths being far less dependent on vision in courtship, once mated, females stop calling with the pheromone, although several tussling males may already have been attracted. Much still remains to be found out about courtship in moths, e.g. whether they indulge in aerial pursuits, as often seen in butterflies.

Attraction of the sexes is only part of the job of reproducing. When a suitable male reaches an unmated female, he has to present himself as friend not foe. He needs to communicate that he belongs to the very same species, not a mistakenly attracted close relative, and recognition may have to be in the dark. Next, he has to be accepted by the female and persuade her not to call to other males anymore. And all this with possibly many other frenzied competitors fluttering around with the same goal! A delicate negotiation is carried out via an exchange of signals and some behavioural fine tuning by either sex. This negotiation phase, like a first date, can be carried out either chemically, acoustically or physically. Male sex pheromones usually act over a much shorter distance than female ones. They can be wafted in front of the female or directly applied to modify her behaviour. Sometimes the males emit a sound (often ultrasound) and may adopt curious postures or perform ritualized dances.

ELABORATION OF MALE AND FEMALE SCENT STRUCTURES

The huge variety of scent structures seen in male moths far outstrips that of females, whose sex pheromone glands are relatively standard in appearance and located on the abdomen, most often between the last two segments, although other positions are possible. Sometimes the female scent organs are not really externally discernible, while in some species, notably females in the superfamily Bombycoidea, the glands appear as more or less elongated, balloon-like protrusions of the body wall. In other groups like tiger moths, they extend far inside the abdomen and open externally via a duct. The quantity of female pheromone necessary can be measured in minute volumes, since just a few molecules can elicit a male response.

The more moth species that are studied, the more types of male scent organs that are discovered, often in bizarre configurations. In their simplest form, male scent organs consist of patches of modified scales known as androconia (singular, androconium). These scales are supplied with male sex pheromone at the base. This supply derives from glands composed of single or multiple cells arranged in bundles. Males of *Peridrome orbicularis*, an erebid moth from the Asian tropics, sport such impressive androconial patches on their wings that the overall wing shape is distorted.

Otherwise scales are arranged on more complex structures, which are usually hidden when not in use. Sometimes the scale bearing organ consists of inflatable tubes, called coremata, equipped with hair-like scales. We have already discussed above under lekking the giant coremata of the tropical genus *Creatonotos*. In temperate regions, curious coremata that smell like lettuce are found in the Ruby Tiger (*Phragmatobia fuliginosa*). The Obscure Wainscot (*Leucania obsoleta*) and Shoulder-striped Wainscot (*L. comma*), have even longer ones tucked into a pouch at the base of the abdomen.

Very often, scent scales are arranged as brushes. When stowed away, the precise arrangement of these organs can be revealed by carefully removing the clothing of covering scales. Knowing exactly how males disseminate their 'Eau de Cologne', exposing their scent organs in full darkness at the time when the sexes actually engage in courtship, is another matter. The best known studies are those involving species of economic importance, such as the Tobacco Budworm (*Chloridea virescens*), since knowing the finest details of a pest's reproductive behaviour can help tailor its control. Some of the most bizarre-looking brushes are those already noted (see p. 32) occurring on the foreleg tibiae of litter moths or erebid fanfoots (subfamily Herminiinae) such as

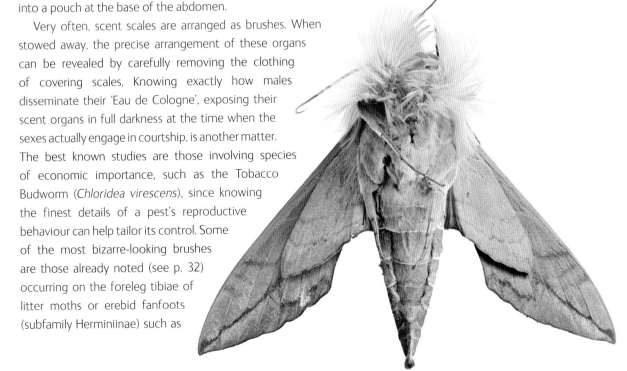

BELOW The fluffy scent brushes of the male Anchemola Sphinx (*Eumorpha anchemolus*) are located at the base of each foreleg coxa. These fans, which can cover half the entire moth, are easily exposed by tugging the tips of the forelegs.

RIGHT The male of *Erebus macrops* has copious and sticky yellow scales that are everted from special pouches on its hindwings.

BELOW The male of the Scarlet-bodied Wasp Moth (*Cosmosoma myrodora*) showers his partner with a confetti of filaments from his abdominal pouches, which cloak the female in a unique kind of protective 'wedding veil'. Such a cloud is known as a flocculent, and contains desirable compounds (pyrrolizidine alkaloids) that protect her against spiders and are daubed on her eggs.

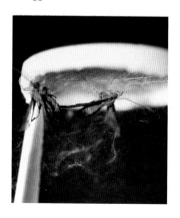

in the genera *Simplicia*, *Renia* and *Macrochilo*. Some of the 'sloth moths' (see p. 62) have an unusual pouch that protrudes into the forewing margin. Particularly large pouches are also seen on the hindwing of *Erebus macrops* and its close relatives. At first glance, these *Erebus* males seem to have a somewhat stunted hindwing compared to that of females. This is actually due to its forward margin that folds over the upper surface of the wing to form a long, stretched pouch filled with sticky yellow scent hair-like scales that are exposed during courtship.

Several species of Sphingidae have impressive paired hair pencils (so called because they are cylindrical before they are fanned out). These hair pencils evert into pompom-like structures normally hidden in lateral slits at the base of the abdomen, which are turned inside out during sexual interaction. Many of the owlet moths (Noctuidae) stow their hair pencils in a similar way to hawkmoths, although these brushes usually hang down from rod-like levers. While at rest in their abdominal pockets, the hair pencils are impregnated with the male pheromone which is derived from glands at base of the abdomen that discharge directly on to the hairs inside the pockets. In other owlet moths, such as *Mythimna sicula*, the levers are rotated horizontally and the hair pencils come into contact directly with bubble-like glands on the venter of the first visible abdominal segment. In other species, the glands consist of single cells grouped together loosely, either at the base of scale, or individually attached to it.

SINGING MOTHS

Moths make far more noises than we give them credit for. At close quarters, soft croons and whispers are employed, most of which are far outside the range of human hearing. These subtle serenades may provide a private communication channel with females unnoticed by predators. The sound producing organs on the thoracic tegulae (tegular tymbals) of the Lesser Wax Moth (*Achroia grisella*) produce very high ultrasound. Clicks on both wing upstroke and downstroke are orchestrated in their wing fanning as a buzz. This noise chimes as a love duet between males and females, yet it is the females that are actually lured.

One moth species is known to employ multiple sound strategies. In a 2014 study of the Yellow Peach Moth (*Conogethes punctiferalis*), Ryo Nakano and collaborators found that males produce alternating sound emissions. One is a long burst of ultrasound capable of eliciting the mate acceptance behaviour in females, and the other consists of a series of short pulses mimicking those of approaching insectivorous bats. The latter is a remarkable ploy to repel competing males. In 2009 Nakano's team also demonstrated in the Asian Corn Borer (*Ostrinia furnacalis*) that these moths could discreetly outfox the attentions of eavesdropping bats by what they romantically called 'private ultrasonic whispering' delivered in close proximity to the female.

LOOKS ALSO MATTER

Butterflies largely rely on visual communication for mate location, whereas moths typically follow a pheromone-mediated system based on female calling (that is long-range) and male 'seduction' (which is short-distance), irrespective of whether the moths have diurnal or nocturnal activity. In many typically day-flying moth families, such as sesiids and zygaenids, the traditional female calling system of moths is used. Visual interaction is likely to be important in several families of primitive moths that swarm by day, displaying their metallic scales at rest or in flight, but often little is known of their pheromone systems. However, there are a few examples where the usual female calling is abandoned. The Ghost Swift (p. 82) is an example of a crepuscular moth where the male colour pattern is visually important in drawing in females.

Victor Sarto i Monteys and his colleagues discovered an interesting case of visual communication in the Palm Tree Borer (*Paysandisia archon*; Castniidae, a family of butterfly-like moths with clubbed antennae). The female does not seem to release

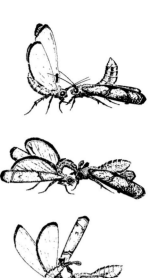

ABOVE Stages in courtship and mating of Cacao Moth (*Ephestia elutella*). After a male (top) approaches a calling female with her abdomen upraised, the pair make contact with their antennae, and then the male curves his abdomen to thump the female's head with his abdominal scent organs everted, and subsequently he performs quick about-turns until he can copulate, finally assuming a back-to-back position.

LEFT Perching Palm Borer Moth (*Paysandisia archon*). While female perfumes reign supreme in moths, this species dispenses with a female one. Sight matters here (as in most butterflies), as males pursue females happening to cross their territories, but the males also smell good to the female.

a long-range pheromone, and instead the male searches for females visually as do butterflies. These moths are conspicuous in flight, as both sexes have brightly coloured hindwings. The males frenetically patrol their beats in bright sunshine, when flying virgin females are also active. *Paysandisia archon* males also resemble some butterflies, as well as many moths, in that they employ short-range pheromones in soliciting females. Thus the overall pattern is as in butterflies, without a female-calling pheromone.

COPULATION

The duration of copulation varies greatly between moth species. Species whose sexes mate only once do not need to engage in a lengthy intercourse, not least because this act exposes them to the risk of predation when they are conjoined. In polygamous species, however, males tend to prolong coupling to prevent other males from reproducing with their partners. Unlike butterflies, a pair of moths rarely takes flight, but if it does it is usually the female that carries the male. Mating behaviour in nocturnal moths in the wild is poorly documented – mating noctuids have rarely been photographed at night even in the commonest species. In looper moths with wingless females (see p. 31), the female is usually the one that supports the weight of a male whilst holding on to a tree trunk.

ABOVE RIGHT A pair of Cecropia Moth (*Hyalophora cecropia*) can mate for more than 24 hours. Their continuously pulsating abdomens display a warning colouration.

RIGHT Among the most spectacular cases of sexual dimorphism in a micromoth is that shown by a mating pair of the Australian psychid *Cebysa leucotelas*.

LEFT Some male moths go to great lengths to reach the female. *Megalophanes stetinensis* (two males illustrated, at top) is a bagworm moth which has an extraordinarily extensive telescopic abdomen (right) which allows the male deep access to the bag containing the female (removed from bag; bottom left).

After copulation, males of polygamous species may prolong clasping the female's genital parts, increasing the chances of fertilizing eggs with his own sperm. In several groups of butterflies, a mating plug called a sphragis (usually external) is secreted over parts of the female abdomen to sabotage other males' chances. Strangely, there is apparently only one reported instance of mating plugs in moths; in 1987 John E. Rawlins cited a West Andean agaristine, *Aucula franclemonti*, whose male leaves a hardened plug, internally blocking the opening to the bursa of the female, as a kind of chastity belt.

FERTILIZATION

As a general rule in biology, the gender that produces the most sex cells (gametes) tends to be the one that competes for mates. Even when a female moth produces thousands of eggs, these always number far less than the sperm cells produced by males. Accordingly, males show an evolutionary trend towards behaving as 'playboys' in order to maximize the chances of fertilizing the scarcer and thus more precious female gametes (eggs). They thus frantically look for females and engage in competition with members of their own sex to get access to a mate. Males are definitely the dispensable sex.

In turn, females may benefit from being promiscuous too. As already discussed, males place their sperm into spermatophores, packages that are transferred in a semi-fluid form into the female genital apparatus. There they harden and assume their final shape, which is often species-characteristic. Spermatophore production represents a big expenditure for males, using up reserves (carbohydrates, fatty

ABOVE A mating pair of Scarlet Tiger (*Callimorpha dominula*); female top, male below. Many female arctiine erebid moths are promiscuous and gain energy and nutrient from the spermatophores of more than one male.

acids, proteins, sterols, phospholipids and defensive compounds). Usually, only one spermatophore per mate can be produced by the male. Accordingly, by counting the number of spermatophores or their residues in a female's bursa, it is possible to assess the number of matings she has had. Frequently it is more than one, and in the most promiscuous species, it can surpass 10. For example, in the Rattlebox Moth (*Utetheisa ornatrix*), 13 have been recorded; in the noctuid *Euxoa perolivalis*, 16. Considering the risk of facing predation or sudden adverse weather conditions, there must be good reasons for delaying oviposition and engaging in such intense sexual activity. Any enjoyment moths experience during this process is that we project anthropomorphically upon them!

In order to illuminate the reproductive process in Lepidoptera, food containing radioactive markers can be administered to male moths and the pathway taken by the various nutrients can then be tracked. Carlos Blanco and colleagues did such work in 2006 on the Tobacco Budworm Moth (*Chloridea virescens*). Radioactive signatures could not only be detected in male spermatophores, but also in the bodies of their mates, as well as in the eggs laid subsequently. It was thereby shown that females may use nutrients and energy obtained from spermatophores to improve their fitness and directly contribute to egg development. In short, having sex with more partners is a crafty way of feeding! The two sexes thus have their own interests in evolving a polygamous behaviour, though for totally different reasons. Research carried out in 2004 by Alexander Bezzerides and collaborators on *Utetheisa ornatrix* has also shown that males of this species transfer protective pyrrolizidine alkaloids (PAs) to females and eggs via spermatophores, and that this helps to prevent attacks by minute *Trichogramma* wasp parasitoids on eggs.

Needless to say, delivering sperm to a female, who is likely to just absorb the male's spermatophore without making ultimate use of his reproductive cells, is a less than favourable bargain – time, nourishment, energy and possibly the only chance of getting access to a female are wasted! To counteract this, males are under selective pressure to develop any possible system that will overcome their competitors' sperm. The contest between males therefore continues inside the female body as a war between sperm from different mates. In many moths, the females bear specialized spines (in structures called signa) on the inside of the copulatory sac (bursa copulatrix). The sharpness of these spines and/or muscular contractions of the bursa can potentially rupture a spermatophore, the female benefitting from nutrients within

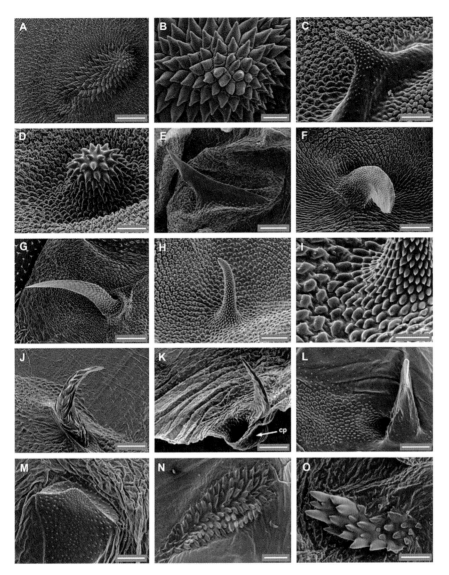

LEFT Tortricidae micromoths have a rich diversity of microstructures on the wall of the female's bursa copulatrix, suggesting that sex in these moths is very complicated.

the ejaculate, which in some cases may comprise 10% of a single male's body weight! Some male moths, in turn, have spines (termed cornuti) on the vesica (see p. 35) which, when deciduous, litter the bursa like a minefield for subsequent competitors, as in the fruit-piercing moth *Eudocima phalonia*. Actually, it is possible that the signa, in some cases, protects the copulatory sac from damage inflicted by the cornuti when they are not shed. The female's reproductive tracts become a battlefield of the genders, the complexity of which remain to be fully understood. Female tortricids in particular have a diversity of remarkable microstructures in the wall of the bursa copulatrix as documented by María Piedad Lincango and colleagues.

Sperm competition can develop in various ways, but an important role seems to be played by a particular type of male gamete cell widespread in the Lepidoptera, called apyrene sperm, extensively reviewed by Robert Silberglied in 1984. Unlike normal sperm, apyrene cells are devoid of genetic material, but both types are stockpiled in spermatophores and thence transferred to females. Though they do not take part in fertilization, some evidence suggests that apyrene cells help to displace sperm from previous males. The competitors sabotage the playground for other males and their rival sperm by filling up the female reproductive ducts, so reducing her receptiveness to other potential mates. Functions of the apyrene sperm in Lepidoptera may, however, be manifold – Michael Friedländer and his colleagues have suggested no less than ten possible roles.

VIRGIN BIRTH

Last but not least, mating can be skipped entirely. The expenditure of energy and time taken to reproduce sexually is very high. Pairs engaging via sex also face costs in terms of increased exposure to climatic extremes or predation, as courtship and mating usually lead to higher visibility and sluggishness. It should not then be surprising that some species have evolved a much more productive strategy of self-perpetuating from one generation to the next, consisting of giving up on sex! These indulge in so-called parthenogenesis. Parthenogenetic moths consist entirely of females which lay fertile eggs, giving rise to other females, propagating in an endless unisexual lifestyle.

To illustrate the potential selective advantages of parthenogenesis, take two females which each, say, lays 100 eggs per generation. One is of a normally sexually reproducing species, the other belongs to a parthenogenetic race. Assume no mortalities occur and all adults reproduce. The initial 100 eggs in the sexual case will produce sexes in approximately equal numbers, say 50 each. The females will then produce 50 x 100 = 5,000 eggs, but half the daughters, that is 2,500. Females of the next and subsequent generations will then number (2,500 x 100)/2 = 125,000 and (125,000 x 100)/2 = 6,250,000, respectively. By contrast, in the parthenogenetic case, the number of females will always match that of the eggs, because no males are produced. Then, for the three generations following the 100 adults deriving from the initial egglaying, a same factor of 100 eggs produced per female will lead to 10,000, a million and 100 million females, respectively. The latter is 16 times the number of reproducing females for the sexual case. Such exponential growth would generate an astronomic difference after just a few more generations. It seems surprising given such higher figures that very few species are parthenogenetic. Crucially with sexual reproduction, however, different gene combinations are produced, so the offspring will be more variable amongst themselves. In the long run then sex pays dividends in terms of exploitation of broader ecological conditions and adaptation to environmental change. In contrast, parthenogenetic species will consist of genetically similar individuals (in extreme cases, they will just be clones). These will be less able to cope with ecologically complex environments and changing climatic scenarios, and populations will be more easily eliminated over time, seldom persisting and most extant examples being relatively recent.

BELOW A wingless female Lichen Case-bearer (*Dahlica lichenella*), goes about egglaying immediately after emerging from her pupa, without mating. This species represents one of the rare cases of virgin birth (parthenogenesis) in moths. In some populations, but not in the UK, winged males occur.

Parthenogenesis is, in any case, a complex phenomenon, as it may occur cyclically in species with alternate sexual and asexual generations (as in aphids). It may also occur haphazardly in some females under particular circumstances (e.g. in normally sexually reproducing species when there are no males around), and most often it is geographically restricted, so that some populations are parthenogenetic and others are not. Despite this, there are scattered reports of the phenomenon across moths.

One family, the bagworm moths (Psychidae), is famous for parthenogenesis. Several species of the genera *Dahlica*, *Apterona*, *Reisseronia* and *Luffia* are known either to be exclusively parthenogenetic or to consist of an ecologically or geographically complex array of populations. Some of these are obligatory parthenogenetic and others sexually reproducing. Things can get even more complicated, as shown by the Virgin Bagworm (*Luffia lapidella*). Here, besides a bisexual species (*lapidella* proper), there is a parthenogenetic race (form *ferchaultella*), sometimes considered sufficiently distinct to deserve species status, and also another all-female race (form *maggiella*). The latter seems to reproduce normally after mating with males, but in reality it only needs the male sperm to activate the eggs, while the sperm degenerates inside the eggs without contributing genetically to the embryo. Parthenogenetic forms of Psychidae are all apterous (wingless) females, and there is a tendency for apterism and parthenogenesis to combine together in other families too, such as in some arctiine Erebidae (*Andesobia*) and in a species of the genus *Mesocelis* in Lasiocampidae, studied by J. Sneyd Taylor in South Africa, whose females oviposit inside the cocoon. It is also found in races of Geometridae (Fall Cankerworm, *Alsophila pometaria*) and lymantriine Erebidae (*Orgyia antiqua*), occasionally also in Heterogynidae.

Other examples of parthenogenetic moths are, however, fully winged and able to fly, like the Coppery Long-horn (*Nemophora cupriacella*; Adelidae), Virgin Pigmy (*Ectoedemia argyropeza*; Nepticulidae) and the leafmining Large Midget (*Phyllonorycter emberizaepenella*; Gracillariidae). The latter was shown by Raimondas Mozūraitis and co-workers to be strictly parthenogenetic, though its females still indulge fruitlessly in calling behaviour with the likely emission of sex pheromone. Winged parthenogenetic forms are also known in Yponomeutidae (*Ypsolopha lucella*) and Gelechiidae (*Tuta absoluta*).

In this chapter, we have seen some of the amazing mechanisms moths use to win over the opposite sex, and ensure, through successful reproduction, that a wide array of gene combinations is passed on to the next generation. Next, we examine the myriad of ways by which moths defend themselves against attack and thus survive to mate.

BELOW The Coppery Long-horn (*Nemophora cupriacella*) is the only member of its genus or even the family Adelidae known to be parthenogenetic and 'males' are either exceedingly rare or represent misidentified of other species.

CHAPTER 5

Moth warfare

I N THIS CHAPTER, WE DISCUSS NATURAL ENEMIES of moths and the ongoing arms race between moths and their adversaries, as moths evolve new strategies for defence while their foes evolve new methods of attack.

ENEMIES

Moths face many enemies in the wild, from bats to parasitoid wasps, but sadly their most important foe today is humans. In some parts of the world, human activities have reduced moth populations drastically (see Chp. 8).

PREDATION

Moth predators include birds, bats, primates, small insectivorous mammals, amphibians and reptiles, as well as a great variety of insects and other arthropods, notably mantises, robber flies, ants, wasps and spiders. Vertebrate predators have a huge impact on moth populations – one pair of Blue Tits (*Parus caeruleus*), can supply their fledgelings with 10,000 caterpillars in a breeding season, and a single Wrinkle-Lipped bat (*Tadarida plicata*), may consume a substantial fraction of its body weight (15 g) in a single night, and a fifth of its diet may be moths. Donald Macfarlane and coauthors calculated that a single cave colony of 275,000 bats could consume 927 tons in a year (possibly equivalent to 185 tons of moths or 2 g per bat per night). Many bats like the Western Barbastelle (*Barbastella barbastellus*) eat almost entirely moths. DCL observed on Barro Colorado Island a single Cane Toad (*Bufo marinus*), voraciously consuming large hawkmoths. Among the insects, Hymenoptera are some of the most important predators of moths. Some of the larger wasps like hornets, and also ants, are major predators of all stages of moths, particularly caterpillars. True bugs (Hemiptera) and beetles (Coleoptera) more rarely devour adult moths, though many attack caterpillars. Bugs essentially stab caterpillars using a sucking stylet to feed on the victims' body liquids, whereas beetles gnaw the caterpillars with their mouthparts. Spiders are major predators of moths, usually using their webs to catch their prey, though sometimes, as with jumping spiders and tarantulas, hunting freely.

OPPOSITE The Jersey Tiger Moth (*Euplagia quadripunctaria*) in one of its spectacular congregations in the Valley of the Butterflies in the island of Rhodes, Greece. During the driest part of the summer, this valley provides a cool shelter for moths, which assemble in the shade among foliage, on barks and rocks, and en masse resemble hanging fern leaves. When disturbed, they display their red hindwings.

RIGHT Feeding on flowers can be hazardous. A Goldenrod Crab Spider (*Misumena vatia*), disguised as part of a yellow composite flower, grabs and devours a Chimney Sweeper moth (*Odezia atrata*).

BELOW The Forest Caterpillar Hunter (*Calosoma sycophanta*) finishing its meal of the larva of a Gypsy Moth (*Lymantria dispar*). The beetle is so dependent on the moth that its population can crash when it has exhausted its prey population.

Certain flower-mimicking crab spiders (Thomisidae), often ambush unsuspecting nectaring moths. Some spiders go to the most incredible lengths to capture their prey via an aggressive mimicry of moth pheromones. With over 70 species occurring worldwide, bolas spiders are a group of the better known orb-weavers that gave up on their typical webs and instead rotate a sticky cord in the manner of gauchos. Such spiders (the American *Mastophora*, African *Cladomelea* and Australasian *Ordgarius* and others more or less related to true bolas spiders such as the Australian *Celaenia*, Neotropical *Taczanowskia* and *Kaira*) have developed the uncanny ability to synthesize common components of specific moth sex pheromones, which they use to lure their unsuspecting male prey. The spiders often combine the lure with a lethal trap of sticky droplets on their silken boleadoras. The capacity to produce chemicals that mimic moth pheromones is also shown by some typical orb-weaving spiders such as the North American *Argiope aurantia*, which is known to lure male ceratocampine and hemileucine silkmoths to their deaths.

Predation is not just a fortuitous event. Sometimes, prey and predator show closely linked cycles of abundance. In southern Europe, outbreaks of caterpillars of the Gypsy Moth (*Lymantria dispar*) defoliate patches of oak woodland, thus boosting the population of their enemy – a tree-climbing carabid beetle, appropriately called the Forest Caterpillar Hunter (*Calosoma sycophanta*), which predates on them both as a larva and adult. A decline in moths due to predation allows the trees to flourish again but then the predator population crashes too because of the shortage of caterpillars. Subsequently, the *Lymantria* population can build up again and the cycle repeats. The impact of the beetles is exceptional – research shows that a pair of adult *Calosoma* and their first generation offspring can destroy over 6,000 Lymantria caterpillars and pupae in a year.

PARASITOIDS: WASPS AND FLIES

Many Hymenoptera plague moths not just by simply devouring them but attacking and finally killing them using a gradual strategy involving a slow death! This strategy is called parasitoidism. Parasitoids are organisms that exploit their hosts, as do parasites, but eventually kill them, as do predators. A spectrum can exist. AZ noted that a sphecid wasp, *Ectemnius kriechbaumeri*, actually captures and paralyzes adult burnet moths, then transports them to its nest of rotten wood, where it provisions each cell with about four individuals. A wasp larva hatches in each of these cells where it devours the paralyzed burnet bodies. This illustrates a strategy intermediate between simple predation and parasitoidism. A vast diversity of parasitoid wasps attack moths. In fact, most Lepidoptera probably have one or more enemies that exploit them at an early stage. Some of these wasps are minute enough to broach eggs, while many attack caterpillars and even pupae. Seeing parasitoid larvae or adults emerge from a caterpillar or pupa while rearing it can be startling. Gruesome as it sounds, internally feeding parasitoid wasp larvae are mostly adept at leaving the caterpillars' vital organs to the very end so they can keep feeding on fresh meat.

Chalcidoidea ranks among the most diverse superfamilies of all Hymenoptera. A single genus, *Trichogramma*, specializes in the eggs of Lepidoptera, using subtle detection cues such as the female moth pheromones or volatile compounds of the hostplant. The tiny wasp drills through the egg shell of its host to lay eggs, and eventually one or more wasps emerge, having completed both larval and pupal stages within. Wasp mating can even occur in the egg. Utako Kurosu studied a lymantriine erebid moth, the Yellow Legged Tussock Moth (*Ivela auripes*) in which, of around ten wasps in a single egg, some matings actually take place inside the egg before the fertilized female wasps fly away. One biologically remarkable chalcidoid parasitoid wasp genus that has been widely used to control owlet moths, whose larvae are significant pests of roots, is *Copidosoma* (Encyrtidae). Wasps in this genus are polyembryonic – they deposit one or two eggs into the host moth's egg and the *Copidosoma* egg then proliferates into multiple embryos, so cloning up to 3,000 offspring.

Another very important superfamily of wasp parasitoids is the Ichneumonoidea of which the Braconidae is one of the most well known families. Many braconid larvae eat their way out of a caterpillar, then spin conspicuous, often white silken cocoons

ABOVE A rare parasitoid wasp (*Sphinctus serotinus*) needs to lay its egg into the very tough cuticle of the final instar of a caterpillar on which it is a specialist, the Festoon (*Apoda limacodes*). As the egg is laid near the surface, earlier instars could slough it off with a skin change.

around an infected individual. Some braconids inject viruses called bracoviruses with their eggs. These viruses have long been incorporated into the genetic makeup of the wasp, and the wasps co-opt them, helping to suppress the caterpillar's immune system. Other main parasitoids of moths, often specializing in leafminers, are the Eulophidae (for example the genus *Euplectrus*). Eulophid wasp grubs, some of which can be surprisingly large compared with their hosts, gorge themselves on the outside of the caterpillar in the mine, sucking it dry.

It is likely that pupal parasitoids, as well as parasitoids of parasitoids (really minute wasps that are known as hyperparasitoids), occasionally run over into the adult moth stage. Surprisingly though, there is no published example of a hymenopteran parasitoid of adult moths, though Francesca Vegliante once observed live wasp larvae inside the abdomen of an *Ancylolomia* species (Crambidae). Michael Boppré tabulated the few known examples of parasitoids of adult Lepidoptera and all were flies. Curiously, two of these instances were maggots of the tachinid *Phryxe vulgaris* in the abdomen of a burnet moth and a geometrid, respectively, while another was of the flesh fly family Sarcophagidae (genus *Arachnidomyia*), whose maggots attacked a shark moth (*Cucullia* sp.). DCL once found a *Urapteritra* moth (Uraniidae) in Madagascar; fly maggots exploded from its abdomen on handling.

The general deficit of reports of parasitoids of adult moths is largely a mystery. The relatively short lifespan of adults does not seem a good explanation, since some live for many weeks. Other factors that explain the lack of parasitoidism of adults include the fact that they are a highly moving target (and therefore difficult to 'pin' down) and that they are usually orders of magnitude less abundant than the other three stages. Also, the adult is not a huge sac of directly absorbable nutrients, as is the caterpillar. In practical terms, it is very likely that the scales of adults pose a near-impenetrable armour, forming a slippery barrier against ovipositing wasps and flies.

Among the flies (Diptera), the key family of moth parasitoids is the Tachinidae, which either inject their eggs into the caterpillar or deposit them on its cuticle, where they are often laid ready to hatch without risk of detachment by moulting. None are small enough to attack eggs. Often they hatch through a hole in a pupa but they do not normally exit from the adult moth. 'Parasite flies' include several important biological control agents for which one of the most catastrophic introductions was that of *Bessa remota*, potentially causing moth extinctions (see p. 177). Another tachinid, *Compsilura concinnata*, introduced into North America in 1909 for control of the Gypsy Moth, also had unintended consequences as the fly attacks hundreds of non-target moth species.

PARASITES

Strictly speaking, true parasites are organisms markedly smaller than the hosts that they exploit and, more crucially, they do not kill their hosts. Among them, mites represent a major group affecting moths. The brilliant red-coloured juveniles of the family

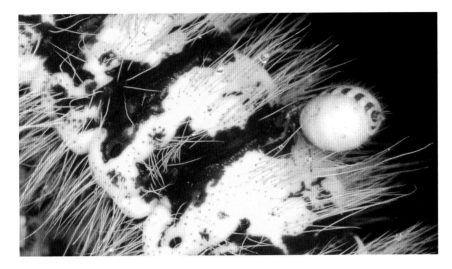

LEFT A female Stick Tick (*Forcipomyia* sp.) sucks the haemolymph of an unidentified caterpillar, probably Notodontidae, in the Madagascan rainforest. These strange organisms are not ticks but flies and are not always harmless as they can allow infections to get in.

Erythraeidae are particularly conspicuous, hanging down from moth antennae, legs or parts of the body where the cuticle is thin. They also attack caterpillars. In fact, these parasites embed their mouthparts into the host's cuticle to feed on its haemolymph. Some mites have evolved to live on specific parts of adult moths. *Blattisocius* mites live in the thorax and *Proctolaelaps* mites are almost always found at the base of the moth's labial palps. Some mites are very large and conspicuous, appearing like bright red or yellow jellybeans. In the case of moths that have hearing organs, some mites can cause deafness if they enter them in large numbers. These mites are decidedly gregarious and nearly always gather in one ear and leave the other uninfected (and fully functional), presumably so the moths can still hear hunting bats.

Moth caterpillars (occasionally adults) may also be attacked by 'stick ticks', which are the females of *Forcipomyia* flies (Ceratopogonidae). The abdomens of stick ticks become so inflated with caterpillar haemolymph they really resemble ticks. These female flies eventually drop off, like satiated leeches, and their puncture marks can eventually kill the caterpillar, possibly by allowing in microbial infection. They can also drink from wing veins of adult moths, and spines on their tarsi may enable them to latch onto a scale. A female *Forcipomyia* has been filmed landing and feeding deliberately on the back of a Brassy Long Horn (*Nemophora metallica*) that was nectaring and ovipositing on Field Scabious.

Finally, moths may sometimes be attacked by nematodes (Nematoda) and hairworms (Nematomorpha), which may sometimes bloat the abdomen of caterpillars, such as those of the Eastern Eggar (*Eriogaster catax*), and Codling Moth (*Cydia pomonella*). It is rare however to witness an unravelling nematomorph 'exploding' from a caterpillar abdomen. Most nematodes which attack moths belong to the family Mermithidae; their eggs hatch in caterpillars which they eventually kill. Interestingly, a nematode, *Noctuidonema guyanensis* (Acugutturidae) that is ectoparasitic on adult moths in French Guiana, even hitches a ride on migrant Fall Armyworm (*Spodoptera frugiperda*) that travel to North America. The nematode spends its life sucking, with a long stylet, from the adult moth and can travel across from one sex to the other during mating.

DISEASES

Caterpillars, pupae and adult moths suffer from several infectious diseases. Caterpillars are often victims of fungi, particularly the genera *Cordyceps* and *Ophiocordyceps* (see Chp. 8), which also sometimes attack adults, and can be recognized by finger-like, often coloured fruiting bodies arising from the insects which ingested their spores. A special form of *Cordyceps*, known as *Beauveria bassiana*, causes muscardine disease in silkworms. Moths are also attacked by many other fungi, notably *Entomophthora*, which can be recognized as a velvety white layer covering the body or by microscopical examination. A microsporidian (*Nosema bombycis*), a type of single-celled parasitic fungus, causes the disease of silkworms known as pebrine, which was studied by the famous French microbiologist Louis Pasteur. During the 19th century, this disease virtually destroyed the silk industry in Europe and silk manufacturers tried to substitute the silkworm with wild silkmoths from the Orient, e.g. Ailanthus Silkmoth (*Samia cynthia*), Chinese Oak Silkmoth (*Antheraea pernyi*) and Japanese Oak Silkmoth (*A. yamamai*).

Finally, viruses are an important source of necrosis (tissue death) in caterpillars. It is common to find a caterpillar as a sad-looking bag hanging down from a stem, and even oozing a black liquid. This is a typical sign of a baculovirus attack. Baculoviruses are often packaged as structures made of stable protein crystals that, once ingested, dissolve in the midgut, releasing the viral DNA. Caterpillars are attacked by a great range of bacteria, the most famous of which is *Bacillus thuringiensis* ('BT'), strains of which are used to control many moth pests.

A phenomenon that eggs and caterpillars never seem to have evolved is the ability to glow in the dark, though there are a few casual reports of luminescent caterpillars probably as a result of bacterial infections. The parasitic nematode *Heterorhabditis bacteriophora* is known to infect the Greater Wax Moth (*Galleria mellonella*) and make it temporarily and visibly glow in the dark (probably due, in turn, to the nematode's endosymbiotic bacteria), as well as turning it pinkish red. Birds avoid these caterpillars, possibly a clever ruse by the parasite. Larvae which were tentatively identified as Great Brocade (*Eurois occulta*) were observed luminescing in 1828 by Benjamin A. Gimmerthal and Jean Baptiste A. de Boisduval made a similar observation, noting several caterpillars of Cabbage Moth (*Mamestra brassicae*) that were glowing on grass stems on a warm June night. Such observations are rare, but it is likely that such larvae also had parasitic (e.g. nematode) infections and that the luminescence ultimately came from bacteria.

Moths have a complex bacterial gut 'flora' (a so-called microbiome), including common gut bacteria like *Pseudomonas* and *Enterobacter*. Some of the most interesting bacteria (*Spiroplasma* and *Wolbachia*) live within the cells and are transmitted from generation to generation (usually through mating, egg formation and oviposition). *Wolbachia*, in particular, is known to have dramatic effects on reproduction, even altering the sex ratio of moths (e.g.

BELOW Moth attacked by *Cordyceps* fungus. It is not uncommon especially in the tropics to come across the dead body of an adult moth attacked by a fungus in its last resting place on vegetation.

its infection feminises the crambid *Ostrinia furnacalis*). Infections are known to prevent chlorophyll being withdrawn from leaves in which leaf miners are feeding (see p. 65). Such bacteria that are not always detrimental to moths are termed endosymbionts (organisms living inside others often to mutual benefit), as opposed to endoparasites. Caterpillars can also pick up bacterial (and fungal or viral) infections by eating spores or particles on the outside of the eggshell for example in the Corn Earworm (*Helicoverpa zea*). As well as by all the microbial pathogens mentioned, caterpillars are also attacked by protozoans (e.g. the genus *Ophyrocystis*), notably gregarines, which infect the gut or cuticle, although they are not much studied.

HITCHING A RIDE

Finding mites on a moth is not always a sign of parasite attack but of phoresy, a relationship where one organism uses another as its carrier. Mites and other tiny wingless arthropods often simply climb aboard moths to get a cheap flight, at limited or no expense to their chaperones. AZ once observed a pseudoscorpion in Spain firmly clamping one of its pincers on to the hindleg of a flying Dark Arches (*Apamea monoglypha*). Such observations are very rare.

ABOVE Luminous caterpillar attacked by a nematode worm (*Heterorhabditis bacteriophora*) that was infected by luminous bacteria.

LEFT The moth *Perola villosipes* (Limacodidae) from Matto Grosso, Brazil with phoretic pseudoscorpions. Aside from their weight the pseudoscorpions are a harmless distraction to the moth. The moth itself is protected by an extraordinary resting posture with raised forelegs to mimic frighteningly a rearing-up tarantula exposing its fangs.

MOTHS FIGHT BACK

Moths have evolved a wide range of methods for fending off their foe, which may be physical, visual, chemical and acoustic, and defence mechanisms occur in all stages of the life cycle. However, these different defence systems should not be regarded as always mutually exclusive as they often overlap. Furthermore, most of the strategies are often implemented through elaborate behaviours. The multifarious phenomenon of mimicry was first described in Lepidoptera.

PHYSICAL DEFENCES

Any life stage can utilize a numerical defence as there is safety in numbers. This is shown in examples mentioned below of masses of eggs, caterpillars, or cocoons, or adults such as Jersey Tiger moths on the Island of Rhodes (p. 96). In some gregarious caterpillars, there can even be a division in role. In siblings of *Malacosoma* eggar moths, there are more sluggish and more active caterpillars, depending on the amount of yolk the first instar caterpillars have available to them as embryos. A spectrum of sluggishness could evolve in the colony as a trade-off between rapid growth and increased exposure to predation.

ABOVE The Eastern Tent Caterpillar (*Malacosoma americanum*) deposits eggs in a bundle around a twig.

EGGS Eggs have a wide range of defences. We mentioned in Chapter 3 (p. 40) how some moths protect their eggs using tufts of hairs from the abdomen. These hairs are obviously defensive. William Beebe discovered a type of palisade construction protecting moth eggs – a Neotropical tortricid, *Tinacrucis patulana*, uses long narrow abdominal scales to construct a deep picket fence around her egg mass, laid on a flat portion of a leaf, forming an effective barrier against marauding ants. In order to escape, the hatchlings form a silken ramp. As in the Eastern Tent Caterpillar, the Gypsy Moth produces a thick cushion of eggs, several layers deep, in this case also protected by a frothy mass that hardens around the eggs called spumaline.

LARVAE AND PUPAE Caterpillars often roll up and drop down or descend on a silken thread when disturbed. The silk is a lifeline to get back to where they were feeding when the danger is over. For some species, like members of the metalmark moths (Choreutidae), living under a silken web offers some level of protection. However, many parasitoids and predators are adept at searching for caterpillars concealed in this way. An interesting strategy then is to use an escape hole. Jadranka Rota and David Wagner found that in a choreutid species, *Brenthia monolychna*, caterpillars readily escaped attack by shooting extremely fast through such 'wormholes' to the opposite leaf surface. Moreover two greatly extended abdominal hairs in contact with the webbing on either side of the leaf enable them to check simultaneously on both sides of the leaf for suspect vibrations. They then perform their Houdini act in the proper direction.

Some caterpillars prepare physical obstacles to protect the pupa from wandering predators like ants. William Toone describes how the lymantriine erebid *Ogoa vitrina* from Madagascar first constructs two disc-like 'rat-guards' further down the twig

LEFT A Pagoda Bagworm (Psychidae) from Ecuador, a larval case with added vegetation parts that is beautifully constructed.

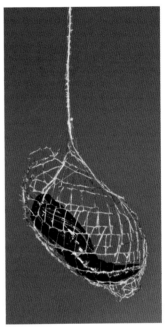

used to support its eventual loose cocoon. Caterpillars of the family Bucculatricidae create some exquisite protective constructions in preparing their cocoon and the larva of Oak Bent-wing (*Bucculatrix ulmella*), builds a palisade of standing silk fascicles that function as pickets to surround its strangely ribbed cocoon.

Some caterpllars use setae, possibly as a physical defence. Uraniid larvae bear conspicuous spoon-tipped bristles, though how exactly these help caterpillars in defence against ants needs proper investigation. The black clavate 'hairs' of the erebid *Tinolius eburneigutta*, which feed on the popular tropical garden plant Black-eyed Susan (*Thunbergia alata*), seem to help in the protection of the larvae against predators. The front two pairs actually move alternately, giving the larvae a comical appearance when on the move. The larvae of the Malagasy moth, *Rhanidophora rhodophora*, are more or less identical in appearance and behaviour. They are also reminiscent of the fully-grown caterpillars of the alder moths (*Acronicta alni* and *A. funeralis*). It has also been thought that these clavate 'hairs' may repel ants, but again their exact function is unknown.

ABOVE The hammock-like network cocoon of a urodid moth (*Urodus* sp.) from Ecuador is stitched like a fisherman's net and suspended on a long line, protecting the pupa inside from marauding ants.

Other larvae seem to exploit ants, likely enhancing their survival. Some uraniids (such as *Alcides agathyrsus*) feed on 'ant trees' (e.g. *Endospermum myrmecophilum*). These trees have vicious ants, *Camponotus quadriceps*, that live in the twigs, and the caterpillars may benefit from reduced attacks by predators. These plants are involved in a mutually beneficial relationship with ants. The trees provide ants with nesting sites (often also with food bodies including extra-floral nectaries on the leaves), while the ants patrol and mop up herbivores, apparently avoiding uraniid caterpillars on the plants. Moths may simply find some benefit in living in proximity to ants. The North American white footman *Crambidia casta*, for example, is known to be loosely associated with *Formica* ants, only living near the edges of nests, where it feeds on lichens. Other moths are in some sense mutualists (when there is a joint benefit for interacting species) with the ants that they appease, although they are not always dependent on specific ants, but may still be protected by them. In 1905, Franz Ebner observed that *Saturnia pyri* caterpillars were attended by small red *Myrmica* ants, which imbibed defensive secretions from

BELOW Egg mass of *Malacosoma* from Georgia, USA, with parasitoid wasp at right hand side. The eggs on the interior are protected by the ones on the outside layer.

the larvae. This may also happen in the Emperor Moth (*S. pavonia*), whose secretions contain proteins, polypeptides and a few aromatic molecules. Saturnia larvae smell like sugary lettuce or rotting fruit. This does not mean that Emperor Moth caterpillars require ants for their development, but they acquire some protective advantage when they are present. Another example of moths that do not actually live in ant nests are the caterpillars of the tortrix moth *Semutophila saccharopa*, which produce anal exudates containing carbohydrates and amino acids, that ants heartily imbibe.

Other caterpillars carry debris on their backs both to disguise themselves and use this as very effective portable armour against predators. In several families, caterpillars build specialized bags or cases a bit like larval caddisflies. Some primitive moths, notably some Incurvariidae and Heliozelidae, cut a protective disc out of a leaf which falls down on a silken lifeline into the leaf litter. Bagworm moths (Psychidae) are the most famous case-bearers and some bagworms have extremely elaborate architectural constructions even resembling pagodas. A few Tineidae, such as the Case-Bearing Clothes Moth (*Tinea pellionella*) and the Hart's-tongue Smut (*Psychoides verhuella*), build larval bags using silk, in the latter example well disguised among fern spores. A number of twirler moth families also build protective cases, notably Lypusidae and some Gelechiidae (*Thiotricha subocellea* constructs a case from the calyces of flowers). Most impressive are the case-bearer moths (Coleophoridae) which build complicated cases out of leaf and twig fragments that they carry around as they feed and, under the protection of this portacabin, they extend their bodies to feed in between the leaf surfaces, leaving a visible perforation in this temporary mine when they move on. Sack bearer moths (Mimallonidae) make cases out of mixed silk, frass, and plant material, out of which they feed externally on leaves. In addition to providing a physical barrier to attack, larval cases also usually assist in camouflage among vegetation or on rocks,

Cocoons, of course, are one of the best lines of physical defence for moth pupae. The structure of the silk cocoon can both protect against desiccation, and, with perforations, prevent waterlogging in very humid environments. In many silkmoths, such as Emperor moths, the cocoon acts as a valve against the entry of predators – the end acts as a tough closed funnel neck that also allows emergence when softened by enzymes spit out by the eclosing adult. Sometimes, as in urodids, the open network cocoon may be on such a long silk line rendering it difficult to tear apart by predators. Limacodids are known as cup moths due to the 'open-lid' appearance of their cocoons when the adult has emerged. In this case, a pre-prepared trapdoor is easy to push outwards in such an egg-shaped structure, but also hard for a predator to find when intact or to breach inwards.

BELOW A White Ermine moth (*Spilosoma lubricipeda*) will 'play possum', appearing dead when disturbed.

ADULTS The first level of defence, especially during daytime when a moth is discovered by a potential enemy, is one of several options. It can drop down relying on good camouflage, or even feign death. The latter ('playing possum') is known as thanatosis, or tonic immobility, when the moth is unresponsive to further stimuli. Usually it rolls on its back and tuck its legs into a 'dead' position, as in the Pale Prominent (*Pterostoma palpina*). Many moths, such as the Red Swordgrass (*Xylena vetusta*), imitate chips of dead bark, as some, like the Horrid Zale Moth (*Zale horrida*), need only stay motionless as their thoracic brushes make them appear long dead from fungal attack. Alternatively, the moth can just fly away. In this case, a moth must balance the need to find a safer resting site with the risk of being chased in flight, sometimes by following a somewhat zigzag route rather than flying straight.

Once a predator picks up a moth, all is not always lost. There is still a chance that the moth can evade imminent death by mechanical countermeasures. Burnet moths, once in a bird's beak, are still resistant to being crushed due to their elasticity allowing them to spring back into shape, and give extra time for their distastefulness to exude through. This is also thought to be a severe form of thanatosis, where the moth is temporarily asphyxiated due to the sudden release of its protective hydrogen cyanide.

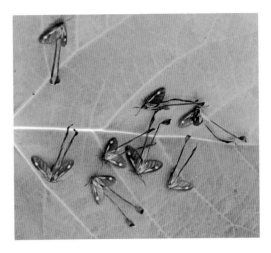

ABOVE If disturbed, *Semioptila spatulipennis* (Himantopteridae) from Congo feign death rather than flying off. In these moths, the function of the long tails has not been investigated.

The entomologist Miriam Rothschild also noted that even severe injuries can clot and heal. Some hawkmoths can jab quite violently with spurs on their legs. According to Ralph W. Flowers, although the Gaudy Sphinx (*Eumorpha labruscae*) very readily uses its needle-like spurs located on its hindlegs, even the frenulum can be used like a sharp, skin-piercing needle as a last defence when grabbed. Both sexes of the hawkmoth *Pseudoclanis grandidieri* in Madagascar, jab with long spurs, at the same time making a loud clattering noise as DCL observed.

Wing shape can sometimes play a vital role in adult defence. A particular mystery has been why some moths, especially males, have very long wing tails. Ths includes the strange adults of the family Himantopteridae and the moon and comet moths (*Actias*, *Argema*). The latter two genera do not only use wing shape (tails) but colour pattern (eyespots) in their defence. The tails in comet moths are usually longest in the male and it was thought that they might help in aerodynamics or bear some special pheromones. Very recently it was discovered that instead, these long tails twirl in flight. The warped paddles at the end of the tails form, in effect, a larger surface whose echolocation footprint draws bat attack away from the body. The Luna Moth (*Actias luna*) can deflect attacks towards the tips of their tails at least 55% of the time, while tailless moths are much more often eaten, even though loss of tails has negligible effect on flight performance.

CHEMICAL DEFENCES

The chemistry of moths is a complex subject and a comprehensive account is well beyond our scope here. There are many ways by which moths can use chemical compounds in their defence. Some are not very obvious to the naked eye. Larvae and adults respond to compounds in the plant substrate and can often detect hostplants by tasting. In the long run, moths have co-evolved with changing plant chemical defences. They can detoxify or excrete powerful compounds of plants and some can denature a wide range of possible plant chemical defences in their guts. They can also accumulate toxic substances. Caterpillars can effectively immunize themselves against their internal enemies. For example, caterpillars can encapsulate or kill parasitoids

ABOVE The Cinnabar Moth (*Tyria jacobaeae*), is one of the most warningly coloured of all European moths and one that is frequently encountered. Its larvae get pyrrolizidine alkaloids from Common Ragwort (*Jacobaea vulgaris*).

internally. They can also fight back against clever wasp larvae that try to sabotage the caterpillar's own immune system. Here we concentrate on striking links between chemicals in moths and their colour patterns, and dramatic ways that caterpillars and adults can use and even eject nasty chemical compounds in their own defence.

TOXICITY Poisons are commonly used by moths to defend themselves against predators, and the most toxin-laden species have a tendency to fly by day, sometimes in addition to flying by night. Like most butterflies, they are often brightly coloured. They have undergone a major shift in their evolution to use more toxic plant families as hostplants. Adoption of poisons in their bodies, either synthesizing them from simpler compounds or concentrating them from their hostplants, has enabled not only adults but caterpillars to flaunt themselves. Any brightly contrasting pattern advertises the moth's repugnant qualities to a would-be predator that spots them by day. Dandy or flamboyant, any moth that successfully gets away from its predator may live to pass on such characteristics. Even if the moth dies in the attack, the lesson learned by the predator (the nasty taste, even an experience of vomiting) can protect other similarly coloured moths, not just the same species, from future predation. Moths advertise this unpalatability with combinations of the 'loudest' colours possible, as seen for example in the gorgeous *Daphoenura fasciata* from Madagascar, although in this and countless other cases, the basis of its toxicity is unknown.

The chemicals utilized by some groups of moths have been investigated extensively. Michael Boppré found that adult tiger moths are attracted to decaying heliotrope plants that contain pyrrolizidine alkaloids and can obtain the compounds from decaying leaves, although others may get them from feeding on flowers. These chemicals are really distasteful to birds and spiders, thus being vital in defence of the moths. Orb-weaving spiders may even cut such bitter pills out of their webs and drop them to the ground, rather than waste injecting them with venom and wrapping them with silk. Caterpillars feeding on ragworts and heliotropes make use of these chemicals too, for example the Cinnabar (*Tyria jacobaeae*) and the Crimson Speckled Footman (*Utetheisa pulchella*), both erebids. Adults of some other toxic moths contain heart poisons (cardenolides). One of the first moths investigated that employed these rather nasty chemicals is the Polka Dot Moth (*Syntomeida epilais*).

As implied above, toxins and warning colours in moths or their caterpillars are only beneficial if predators do not only learn from their errors but go on to avoid similarly advertised potential prey. Moths continually educate their predators. Hence the evolution of mimicry, an important concept in defence that we discuss later.

REFLEX BLEEDING, PROJECTILE SPITTING, VOMITING Many caterpillars regurgitate their gut contents when attacked and some of the compounds can be quite repugnant to predators. Caterpillars feeding on Apiaceae (umbels) and Rutaceae (citruses) vomit droplets rich in fumarocoumarins, which are photo-active compounds that bind to DNA. As is well known when handling Rue (*Ruta graveolens*), these compounds can blister the skin when activated by light. The caterpillars, meanwhile, have their own special system to detoxify the compounds. Vomiting has a cost, especially in caterpillars that stay in close proximity to each other, because vomiting can quickly spread disease. Caterpillars can also actively jet out fluid. The larva of the Puss Moth (*Cerura vinula*) has a gland in its thorax that can be used to aim a jet of formic acid at its provoker. Possibly the red filaments that can be everted from its tails are just a warning that it is willing to use more vicious but costly weapons in its arsenal. The adult moth can also exude yellow droplets from the thorax when freshly hatched.

Many moths exhibit a more extreme form of 'reflex bleeding', they produce yellowish droplets or even chains of foam when picked up. One example is the Faithful Beauty Moth (*Composia fidelissima*) – the foam may also make a series of clicks as the bubbles burst. Such moths often self-mimic these droplets near their head area or on the body and base of the forewing to avoid wasting precious fluids. Chalcosiine zygaenid moths (e.g. the genus *Eterusia*) and Burnet moths, species of *Zygaena*, produce yellow droplets containing haemolymph mixed with cyanide binding substances (cyanogenic glucosides). They can then release hydrogen cyanide after mixing such compounds, normally stored in special cavities inside their cuticle, with suitable enzymes. Using other enzymes, these zygaenids are also capable of inactivating cyanide-containing compounds and are quite resistant to cyanide that would normally arrest their breathing (see p. 106).

Some adult moths just squirt in self-defence. If disturbed soon after emerging, many jet copious pinkish coloured meconium from the anus. This is the waste liquid accumulated during the pupal stage, and a familiar experience to any moth breeder. Tom Eisner recounts that he was once sprayed in the face while in bed and again the following morning as he went out of the door. The culprit was a noctuid, the Army Cutworm Moth (*Euxoa auxiliaris*), which sprays fluid from its rectal sacs when disturbed, each sac ejecting about 5% of the moth's body mass. While this spray is relatively innocuous, sometimes moth spray can be offensive-smelling, as during the aestivation (summer equivalent of hibernation) of the Bogong Moth (*Agrotis infusa*) in Australia. Their spray becomes increasingly offensive towards the end of the dry season.

BELOW The spectacular larva of Drury's Jewel (*Cyclosia papilionaris*) from China secretes clear droplets containing defensive compounds at the end of each tubercle.

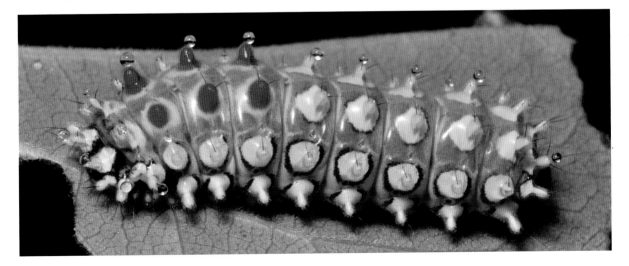

ABOVE *Scopelodes unicolor* (Limacodidae) from Borneo sporting a funky head crest, like a hat, curls up its abdomen in a defensive warning posture.

When handled, many poisonous moths such as *Asota borbonica* (Erebidae: Aganainae) produce an intense, rather unpleasant, green bean-like smell, a signature scent of compounds called pyrazines that have been shown to make predators more alert to other compounds. In most cases, however, moths are often sweet smelling to humans. Some of the odours they produce help to disguise the moth. For instance, it is said that the adult of the Lappet (*Gastropacha quercifolia*) smells rather sweetly of dry oak or beech leaves. Not many moths have a distinctive smell that lasts after death, although it was noticed that specimens of *Terastia* (Crambidae) still had a distinctly unpleasant smell several years after being collected. Cyril L. Collenette has referred to members of the related genus *Agathodes* as 'pig-sty' moths because of this stink, reminiscent to him of Fenugreek seed or curry powder. How these adaptations are used against predators is not known. However, in a vivid portrayal of 100 caterpillars from Costa Rica, Daniel Janzen and colleagues relate that both sexes of the saturniid *Periphoba arcaei* emit a strong garlic-musky odour when picked up, and curl the abdomen revealing warning rings between the segments. When placed in a raiding column of the devastating Burchell's Army Ant (*Eciton burchelli*) the odour cleared an 8 cm (3 in) ring around the motionless moth.

Scales too can be important in defence. We mentioned on p. 88 'flocculents' produced in courtship to overpower a female. Producing fluff can similarly act as a choking defence against predators, a bit like the 'smoke screens' of ink emitted by squid. Species of the ctenuchine erebid genus *Homoeocera* release a candyfloss-like substance that acts as a sort of 'chaff' to disrupt the accuracy of bat sonar. In the case of a female *Pseudarbela* (Psychidae) from Borneo, DCL was astonished to observe the rapid filling of a field lab with particles of fluff when he picked up the moth.

OUTSMARTING PLANT DEFENCES Caterpillars have some ingenious ways to counteract defences of their hostplants. Most plant families have evolved their own unique chemical signatures, a cocktail of sometimes quite toxic substances that can occur inside cells, fill specialized ducts (e.g. resin and latex canals) or are mobilized on demand. That is why many families of plants form effective barriers to exploitation by most groups of moths. Only moth species equipped with particular antidotes such as detoxifying enzymes are able to circumvent these defences. Plant defences can also be mechanical, in that plants with sticky latex can actually gum up the delicate mouthparts of caterpillars. Some plants such as those in the spurge family (Euphorbiaceae) produce a defensive latex in this way to deter being munched on, but some caterpillars have found a way round this. Caterpillars in the genera *Urania* and *Lyssa* bite the main midrib of the leaf or the stalk of an *Omphalea* plant, until the leaf hangs down, so that they can feed unimpeded with the main latex artery severed. This biting strategy is called 'trenching'. The caterpillars of the prominent moth *Theroa zethus* etch the latex veins of other euphorb hostplants such as Poinsettia (*Euphorbia pulcherrima*) by applying an acid secreted

BELOW *Theroa zethus* feeds on Poinsettia (*Euphorbia pulcherrima*), a plant with thick white latex. To defeat this defence, which could gum up its mouthparts and deliver toxins, the later instar caterpillar bites the leaf midrib and delivers an acid from a gland to stem the flow.

from a ventral thoracic gland thus hampering the flow of viscous latex. When plants have spines laden with neurotoxins, caterpillars must avoid being hooked or spiked. They do this by spinning a silk web over the spines. Other plants have mechanical defences such as silica-like shards as seen in many grasses. Caterpillars feeding on such plants have very tough mandibles.

VISUAL DEFENCES

Visual defences concern startling and deflective displays (e.g. fake eyespots), camouflage (cryptic resemblance of different plant parts or other substrates), disruptive colouration, and mimicry of other organisms, usually other insects, which often involves shared striking colours. The latter advertise distastefulness coming from toxins in the body or otherise signal unprofitability as prey.

CAMOUFLAGE Camouflage (or crypsis) is probably the most common defence mechanism for caterpillars and resting adult moths. The degree of resemblance to living or dead plant parts can in many cases be spectacular and neatly reflects the usual environment (e.g. leaf litter) around where they live. There are many examples where moths have transparent windows in their wings, like perforated leaves, or metallic reflective spots that imitate dew. Some moths even sport white spots that resemble fungi as seen on the legs of the overwintering Herald (*Scoliopteryx libatrix*). In species of the South American genus *Neorcarnegia* (Saturniidae), the transparent areas are enlarged to such an extent that only the veins remain visible over large parts of the wings, thus conferring on the moths the texture of leaf skeletons on the forest floor. Being 'leafy' often requires precise imitation of common leaf features. This phenomenon can easily be observed in moths resting with fully spread wings where a single line runs between the forewing tips and crosses the hindwings, thus closely resembling the

LEFT A trio of the dead leaf moth *Oxytenis modestia* from the leaf litter of the Mindo cloud forest, Ecuador. Notice how variable the moths are and how a leaf midrib is imitated by a confluent pattern between the forewing and the hindwing.

midrib of a leaf blade. Moths in the genera *Hamodes, Hypopyra, Lonomia, Oxytenis* and *Sarcinodes* display this effectively. Also, perhaps not surprisingly, the forewing in these moths often has its apex tip drawn out like a leaf stalk.

Phenomena such as these fake leaf midribs reveal the painstaking work of natural selection in realizing an integrative pattern from the pupal wing buds that are independent between forewing and hindwing. Even the slightest mismatch would suppress an adaptively valid visual effect. Spectacular examples of integrative patterns, which often also affect the thorax and the abdomen, are in the families Crambidae (genera *Diaphania* and *Dichocrocis*), Mimallonidae (*Tolypida amaryllis*, in side view), Geometridae (*Agathia, Celenna, Comostola*), and Drepanidae (*Drapetodes* and *Phalacra*). Such rearrangements are particularly intriguing as evolution often makes savings and retains elements of ancestral patterns to combine them with new ones. Males of the leaf-mimicking, fruit-piercing genus *Eudocima* provide an excellent example. These moths rest with wings closely pressed to the body in a roof-like position, and it is the upperside of the forewings only that resembles leaves in this case. Some species still show a bent original main line running from front to hind margin of the forewing (e.g. *E. cajeta* and *E. homaena*), whereas in others (*E. aurantia*

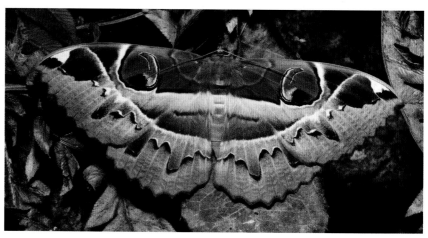

ABOVE RIGHT A crambid moth (*Glyphodes ernalis*) from China provides an example of a disruptive pattern integrated between the forewings and hindwings that tricks a predator into not recognizing its true shape.

RIGHT An erebid moth (*Erebus ephesperis*) from Yunnan, China appears to have a tatty wing outline overlaid over a perfect one. The strange three dimensional effect is achieved by countershading using shadowing that works only when light is cast from above.

LEFT The European geometrid moth (*Apochima flabellaria*) adopts a strange resting posture that camouflages it by rolling its wings and holding its hindwings and forewings at very different angles. Above on the twig are its freshly laid eggs.

and *E. sikhimensis*) only the inferior two thirds of this line survive and merge with a substitution segment reaching the apex. In this way the midrib of a leaf is perfectly reproduced! A few species show intermediate conditions, either the apical segment fails to precisely line up with the main line, or it is weak and the anterior part of the main line has not fully disappeared (as seen in *E. iridescens*, *E. cocalus* and *E. phalonia*).

Revolutionary enhancements in camouflage are seen in many families, notably Thyrididae, Tortricidae, Drepanidae, Geometridae, Notodontidae and Uraniidae. The Epipleminae, a subfamily of the latter, and the geometrid *Apochima flabellaria* have flexion lines along the wings that allow an astonishing degree of wing rolling that increases the moth's resemblance to twigs. Most of the natural features imitated by moths correspond to parts of plants or rock backgrounds, though there are also other (more ephemeral) features such as bird droppings (discussed below), which many moths in the families Tortricidae, Drepanidae and Nolidae closely resemble.

Although moths achieve exceptional disguises, there are constraints that cannot be overcome. One is having a three-dimensional body, which is something no organism can give up. However, evolution has conjured a series of tricks to thwart perception of three-dimensionality by predators. A common illusion is avoiding or reducing the organism's shade so as to appear flat. This effect can either be achieved with feathering rows of hairs outlining the body edge that disguise any gap with the background on which the moth rests, e.g. in *Soxestra* adults or *Gastropacha* caterpillars, and/or by tightly adhering to it, e.g. *Catocala* caterpillars.

Another important method that many moths exploit to look flat is countershading. Consider a cylindrical, uniformly green-coloured caterpillar resting on a leaf. When light hits the body, its dorsal part looks brighter than the sides because of the greater light reflection from above. The *chiaroscuro* effect thus produced enhances conspicuousness of the caterpillar by revealing its three-dimensional body better, which makes it easier for predators to spot their prey. To counter this, many larvae

BELOW One of the most peculiar examples of mimicking plant parts is the caterpillar of the nolid moth *Carea varipes*, whose thorax is spherically inflated to look like a plant gall.

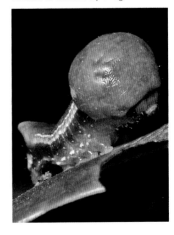

Uropyia meticulodina
(Notodontidae) in China embodies one
of the most breathtaking examples of
countershading in nature to resemble
a dry leaf curved at both margins. With
light shining from above, a gradation
of dark shading at the lower margins
of both light wing areas renders a
convincingly three dimensional optical
illusion (top). The effect completely
disappears if the same moth is turned
on its back, revealing the wing area to
be flat (bottom).

Prismosticta fenestrata
(Endromidae) combines two strategies
to look like a rolled leaf. It gives the
3D illusion of leaf rolling at its forward
forewing edge (top panel), while
physically folding the trailing margin
of its hindwings to match (top panel).
This mirroring leaf-edge effect only
works with light from the top, and
disappears when the moth is oriented
upwards (bottom panel). Note also
that the moth disrupts its bilaterally
symmetric appearance by leaning the
abdomen on one side.

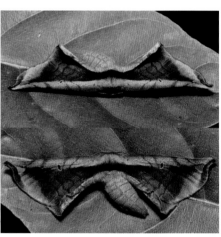

have developed a darker dorsal colouration so that under sunlight the darker green
of the back matches the sides more in shadow, thus obscuring recognition of its
three dimensions; indeed, it may thus look more like a flat leaf.

Other examples exist of making a two-dimensional surface appear three-
dimensional, e.g. like a curled leaf. The adult of the tropical Asian prominent moth
Uropyia meticulodina (Notodontidae) uses a trick of optical perception whereby
pale bands along forewing margins appear to be curled so that the rolled margins
of a dry leaf seem to be airbrushed onto its flat wings. The effect, visible also in
the Green Fruit-piercing Moth (*Eudocima salaminia*) and the endromid *Prismosticta
fenestrata*, is spectacular. Furthermore, the pale bands in these moths also show the
countershading effect. Another almost perfect three-dimensional optical illusion is
shown by *Paralebeda plagifera*, a member of the prominents, which looks like a rolled
cigar at rest. In this case darker slivers of the forewing pattern create the illusion
of layers. The snout moth *Arbinia todilla* also mimics a rather flatter leaf, still with
the margins curled over, again using the same principle and upturned palps which
imitate a leaf-stalk.

The camouflage patterns of adult moths are nothing short of astonishing. One
unidentified Neotropical pyralid (subfamily Chrysauginae) mimics mosses growing on
tree trunks with perfectly raised green scales on its back. Chrysauginae include moths
with the most bizarre resting postures. Species of the genus *Pachypodistes* (p. 24)
probably win any prize for the funkiest boots, and a far cry from any typical moth
posture. Indeed, an important way a moth can camouflage itself is by 'disruption',
where its natural appearance is distorted with eccentric lines, patterns, body parts and
postures that make it hard for a predator to recognize a standard shape. A distinctive
feature of most animals, moths included, is their bilateral symmetry and distorting
this can help to hide them, e.g. in footman moths and *Hyperchiria* saturniids that rest
with one forewing overlapping another.

Moths can also partially achieve this symmetry violation behaviourally by
moving their legs and antennae into an asymmetrical position, as often seen in
attacine saturniids, or skewing the abdomen relative to wings, as do many species

of Mimallonidae and Bombycidae. Some moths like the Lappet (*Gastropacha quercifolia*) and Poplar Hawk-moth (*Laothoe populi*), retain symmetry but they disrupt the expected moth appearance by jutting out the hindwings in front of the forewings at rest. Caterpillars sometimes assume a J-posture, but the disruptive effect of such shapes seems otherwise limited. Truly disruptive patterns, in contrast, help to deflect attention from the body of caterpillars and adults using optical tricks, including diverting diagonal stripes, dots and and false outlines. Amazing examples can be found in adult Crambidae (e.g. *Conchylodes nolckenialis*, most species of *Glyphodes*), Geometridae (e.g. *Pityeia histrionaria*, *Plutodes* spp.), Noctuidae (*Attatha regalis*), various attacine saturniids with translucent windows on their wings (e.g. *Rothschildia* spp.) and many arctiine erebids – the latter often combining these optical tricks with other defence strategies. Many caterpillars also show disruptive patterns, such as those of the American noctuid *Phosphila turbulenta*, whose numerous thin longitudinal black and white stripes potentially confuse predators. In other species there are dots or separate complex pattern elements that break down their cylindrical appearance, as in several shark moth caterpillars (*Shargacucullia*, Noctuidae).

ABOVE One of the most bizarre resting postures in the moth world is that of euteliid moths such as this *Anigraea* species from China, whose abdomen is asymmetrically raised to perfect its disguise as a bit of dead vegetation.

An opposite strategy to diverting attention is attracting it. So called false head mimicry redirects attacks towards fake vital organs, typically with false heads portrayed as far as possible away from the real head. When attacked, the moth then flies away, leaving just a piece of wing in the predator's mouth. Amongst many species using this strategy, few can possibly outperform the false head of the Neotropical erebid *Eulepidotis hermura*. When it is at rest, markings on the outer corners of both wing pairs combine into a false head bearing antennae, with imitation forelegs at the sides. Most importantly, the variously coloured scales are configured in a way that different hues gradually shift one into another, giving a full three-dimensional impression of a thick, velvety moth head rivalling the true one. And, as a last touch of art, scales darker than the orange ground colour simulate an apparent shadow of forelegs and antennae apparently raised off the ground. An example of a false head in micromoths is in the newly discovered family Tonzidae (genus *Tonza*), where the wings drape over the body and an eyespot at the wing tip combines with antennae that are longer than the wings themselves, reinforcing the false head.

False head mimics can sometime look surprisingly similar in species from different continents. Such wing pattern convergence is largely a matter of independent origin of similar solutions in response to comparable selective pressures, rather than as a result of common ancestry. Such is the case for similarities between some Indo-Australian Uraniidae (e.g. *Urapteroides astheniata*), Old-World swallowtail moths (*Ourapteryx*, Geometridae) and a number of *Therinia* (Saturniidae) from the neotropics. At rest these are all similarly-sized white moths with cream or buff-coloured diagonal lines that converge in false head-like hindwing tails. False head

ABOVE The strange colourful dayflying tortricid moth (*Cerace xanthocosma*) from Taiwan, China turns around immediately on landing on a leaf, making its false head appear like the real one.

ABOVE RIGHT Caterpillars of Wavy-lined Emerald Moth (*Synchlora aerata*) have the remarkable ability to use fragments of any flowers or plant they are feeding on to camouflage themselves.

mimicry can also be behaviourally reinforced in day-flying species. Asiatic tortricid moths in the tribe Ceracini have a psychedelic pattern of short white lines on black outlining the forewing margins, enclosing a large area of orange, which is variously spotted. DCL noticed in Taiwan, China that as these moths land after flight, they briefly scuttle on a leaf like a cockroach and then do an about turn, potentially confusing a predator as to their rear head end. When a moth is targeted by a predator, every second counts.

Some moths even have what would usually be considered developmental abnormalities embedded as cut-outs into the wing margin, such as in the Notch-wing Button (*Acleris emargana*). The apparently tattered wing breaks up a predictable moth-like outline against whatever search image a predator expects to encounter. Some moths make a beautifully artistic pattern when grouped together, such as the collectively fern-like patterns of Jersey Tiger Moth (*Euplagia quadripunctaria*), in the 'Valley of the Butterflies' in Petaloudes, Rhodes (p. 96). This behaviour illustrates

RIGHT The larva of *Mustilia* sp. (Endromidae) from China is one of the more bizarre snake mimics. When it is disturbed, as here, it retracts its head, while displaying the colourful areas at the sides of the head to imitate eyes and whilst fanning its side flanges to good effect.

the importance of dual signalling: disguised at a distance in the environment and sometimes advertising warning colours when approached or when disturbed. The Jersey Tiger, however, goes even further towards a triple signalling as, at a short distance, the appearance of resting individuals is greatly disrupted by sharp white bands standing against a black ground colour.

A number of species benefit by being especially convincing as unpleasant, even putrid-smelling, objects. Looking like bird droppings is probably an excellent means of defence, against birds in particular. It is often the mimics of bird droppings that seem to be left over on the outside of a moth trap in cases where birds have already breakfasted on the other moths. The phenomenon occurs widely in moth larvae (which often rest in a suitably contorted position such as that of Alder Moth (*Acronicta alni*), as well as in pupae and adults. Vast numbers of species employ this strategy as adults, notably amongst Tortricidae, Nolidae, Drepanidae (a good example is the Chinese Character, *Cilix glaucata*) and the noctuid genus *Acontia*, and impressively in the Indo-Australian genus *Tonica* (Depressariidae). One Southeast Asian drepanid escalates the bird dropping display shown by some other members of its genus to a whole new level. A species recorded as *Macrocilix maia*, though possibly a different one according to DNA analysis, enacts a disgusting scene, with a large red-eyed fly vividly portrayed

ABOVE LEFT The xyloryctid micromoth (*Tonica* sp.) from Australia is an impressive bird-dropping mimic.

LEFT The remarkable drepanid moth *Macrocilix maia* mimics a red-eyed fly feeding on bird poop, or possibly vomiting. This must be the best known example of apparently near photographic level artwork on a moth's wings, but the scene not only looks disgusting, the moth smells of urea.

on each forewing apparently feeding on (if not vomiting) a splat of brown fluids on the hindwing. These pretend flies have a clearly defined body, legs and even wing reflections. Not only that, the moth takes its depiction of a repugnant reality even further, as it has been reported to have a foetid, urea-like smell. These wing patterns are some of the most impressive visual imitations of any group of organisms that can be found in nature.

DISPLAYS THAT STARTLE As discussed above, the patterning on adult moths and caterpillars often allows them to hide from predators. However, some species use a form of deception that is more combative. Instead of trying to vanish from view, they aim to frighten potential predators. One of the most bizarre displays that confuses or deters predators is enacted by a caterpillar that, instead of rejecting its head capsule at each moult, builds up a whole series of them on its head. This gum skeletonising caterpillar, *Uraba lugens* (Nolidae), has been nicknamed the 'Mad Hatterpillar'. The larvae moult up to 13 times and the resulting stack of heads balanced above its own must look intimidating to a potential predator. If provoked, the caterpillar can thrash the hat stack about in a menacing fashion. In this way they manage to thwart predation, by a marauding shield bug, for example, very effectively.

More familiar are the offputting displays of many moths using at least a pair of false eyes. The eyespots on caterpillars can be extraordinary. For example, the eyespots of *Hemeroplanes triptolemus* seem sculpted in three dimensions with reflective pupils, resembling a snake's eyes. *Hemeroplanes triptolemus* is a striking mimic in particular of the Eyelash Pit Viper (*Bothriechis schlegelii*).

Many hawkmoth larvae exhibit some kind of snake mimicry, using exaggerated eyespots on the thorax and puffing up segments behind their heads. Prominent eyespots feature in other families too, even if they do not qualify as snake mimicry.

RIGHT The Mad Hatterpillar (*Uraba lugens*) builds up head capsules as it grows, giving it this comical appearance. The caterpillar thrashes its head dressing to scare off predators.

Eudocima caterpillars (Erebidae) have particularly striking eyespots on the thorax and abdomen, enhanced in a menacing fashion when they hunch up and raise their tail as if they are about to strike. The banded caterpillar of the Frangipani Sphinx (*Pseudosphinx tetrio*), was supposed by Dan Jansen to mimic coral snakes in its banding. Some caterpillars even change their defence mode during their life. The larva of *Oxytenis naenia* (Saturniidae), resting J-shaped, resembles a bird dropping in the first four instars, but in the fifth it looks like a dried up leaf with spots resembling holes. When disturbed, the caterpillar flashes eyespots under its thorax and when further provoked, pumps up flaps in the thoracic region that reveal snake-like eyespots.

RIGHT The adult of the giant Bentwing Ghost Moth (*Zelotypia stacyi*) in Australia is secretive and rarely spotted by day after its emergence from *Eucalyptus* trunks. It conveys the impression of a snake or alligator enhanced by an eyespot that seems moulded, standing out in relief on its forewings.

The larvae of *Coronidia orithea* (Sematuridae) have two rows of raised black pinhead-shaped warts on two thoracic segments, and a pair for good measure on the eighth abdominal segment, making them strangely reminiscent of the multiple eyes of jumping spider. The larvae of *Sematura lunus* and *Homidiana pictus* also have wart-like structures which may be secretory or somehow make the caterpillars look more distasteful. Frightening eyespots can also be found in pupae. Pupae of species of *Dysphania* (Geometridae) that are wrapped in leaf rolls have two black eyespots to scare any predator that would try to winkle them out.

The flashing of eyespots is a more common behaviour when moths are disturbed. It is seen in many sphingids such as the Eyed Hawk-moth (*Smerinthus ocellata*) and in numerous saturniid genera, above all *Lobobunaea*, *Bunaea*, *Pseudobunaea*, *Automeris* and *Leucanella*. In fact, eyespots are particularly common in saturniids, which have developed endless optical tricks to make them appear like wide open startling eyes of owls or other vertebrates. These tricks aim at circumventing reality, making flat designs appear as convex, tear-lubricated eyes. Just as artists dot a little white patch on eyes to imitate reflected light, saturniids recreate such round pupils and emphasize them using thin white peripheral rings or crescents as if with open eyelids.

There is convincing evidence that a sudden display of eyespots or other vividly coloured patterns on hindwings, when the moth raises its otherwise camouflaged forewings, has the effect of scaring predators, at least inexperienced ones. Such displays may take place when the moth is resting on its perch or through a series of more complex wing movements and convulsive dances on the ground. For example, *Hyles* hawkmoths perform a dance that sometimes incorporates pseudo-stinging motions of the abdomen. More debated is the role of permanently exhibited eyespots such as those that occur on the forewings of the Comet Moth (*Argema mittrei*). It is thought that in some cases they may help deflect the attack of predators away from more vulnerable parts. *Argema mittrei* is unusual compared to *Argema* elsewhere, in that the pupils are small and black, just as in lemur eyes.

Perhaps lemurs are especially mesmerized. It is often possible to find a specimen that got away, with a tattered gap in its wing as evidence of a bird or gecko strike (or 'lemur grip'), around its missing eyespot.

The way that species of *Catocala* exhibit their vividly coloured hindwings when disturbed is quite sensational. These moths typically rest by day on tree trunks with their hindwings fully concealed under the camouflaged forewings but, if disturbed, they suddenly take flight in a frantic zig-zag fashion. This creates a visual strobe effect of coloured flashes that renders their trajectory untrackable and disorientates attackers. The moths then quickly disappear alighting on another tree trunk, usually on the side away from danger, and immediately hide their bold-coloured hindwings beneath the bark-like forewings. This escape flight has low-frequency wing strokes, which contrasts with their normal undisturbed flight at night, when these moths approach food sources in a straight path with a high beat frequency. The adaptive explanation for the difference in beat frequency is that with the more rapid wing strokes during normal flight, the flashing colouration of the hindwings would appear blurred to predators compared to the low beat frequency during the escape flight, when it is clearly exhibited. Even when they are disturbed at night during their normal flight, they suddenly switch to the daytime-characteristic escape flight!

MYSTERIOUS COLOUR PATTERNS There is still much that is not understood in the astonishing diversity of moth wing patterns, and many are bizarre enough that they are unique and not replicated anywhere in the world.

LEFT Members of the genus *Sinna* (Nolidae; this one from Malaysia) sport some of the most bizarre yet exquisite colour patterns among all noctuoid moths (other examples are *Baorisa* and *Mazuca*).

ABOVE A lycid beetle (*Calopteron* sp., left) and an erebid moth (*Correbidia* sp.) are Müllerian co-mimics that closely resemble each other in the French Guianan rainforest. Both are rather soft and distasteful and likely to resist a predation attempt.

As much as patterns can be striking, they can also be hidden. Apart from patterns visible to humans, another area of moth biology that is scarcely understood involves signals or wing patterns in bands of the electromagnetic spectrum not normally detected by humans. Hidden ultraviolet (UV) reflectance patches in males were first made famous in butterflies by Yuri Nekrutenko's studies on the Brimstone (*Gonepteryx rhamni*). But UV reflectance may be a more general phenomenon, not restricted to males. It turns out that many moths have highly UV-reflective scales. Surprisingly, in a study of 883 European Lepidoptera, UV reflectance proved to be much more common in night-flying than day-flying species. Many moth caterpillars are also highly UV reflective at night. The implication is that at night, there is no risk associated with being UV reflective, whereas during the day, UV reflectance poses considerable risk, since birds are sensitive to UV light. Perhaps this is why, in some pierid butterflies, the UV reflective areas are limited to patches in males only, and these are on also the upperside, so there is no risk when at rest with wings closed. Notably, when the unpalatable Gypsy Moth (*Lymantria dispar*) was painted with UV reflective powder, it was more predated on by native birds than those painted with UV absorptive dye. Naturally UV-reflective moths would normally stand out while resting by day against a non-reflective background, so it may be that the forewings tend to be non-UV reflectant whereas the hindwings they cover can remain UV reflectant. Studies on the importance of UV reflectance in moths is ongoing, but the case in moths appears to differ very much from that of discrete upperside patches in some butterflies.

TRUE MIMICRY

Strictly speaking, when organisms imitate others or any part of the surrounding environment in order not to be recognized, they enact camouflage, a phenomenon that usually relies on colouring, shape and behaviour (a kind of passive mimicry of their resting background). In contrast, when true mimicry is involved, it is in the organisms' interest to be spotted by their potential enemies. In fact, their strategy is to be mistaken by the predators for other organisms that happen to be well protected by toxins or other weaponry. It is therefore necessary for inexperienced predators to learn about the harmful or distasteful features of the protected species, so that the subsequent encounters with any organisms displaying a similar colour pattern will result in an avoidance reaction.

There are two classic modes of true mimicry, so called Batesian and Müllerian. In the first instance, which takes its name from the great naturalist Henry Walter Bates who first discovered it through his observations of butterflies in the Amazon, the mimic is palatable to predators but it gains protection by resembling toxic or otherwise well-defended species (termed models). One option is to mimic stinging insects such as wasps and bees. There are some excellent mimics of Hymenoptera in the families Sesiidae (clearwings), Erebidae and Sphingidae. The mimicry can be so convincing that some species of the first two families exhibit a clear 'wasp waist' as well as wasp-like behaviour in flight, including sometimes trailing hindlegs or long scale tufts to mimic a long ovipositor. Many sesiids seem to be rather generalized Hymenoptera mimics combining features of wasps and bees, e.g. looking like wasps in general form but recalling the pollen baskets of bees with fluffy yellow-orange scales on the hindlegs (e.g. *Melittia*). Alternatively, mimicry can be very specific as in the Poplar Hornet Clearwing (*Sesia apiformis*), which is a near perfect mimic of the Hornet (*Vespa crabro*). Strongly bee-mimicking moths include the day-flying hawkmoths of the genera *Hemaris*, *Cephonodes* (resembling bumblebees) and *Sataspes*

LEFT An exceptionally good case of Neotropical wasp mimicry shown between a *Polybia* wasp (left) and the sesiid moth *Pseudosphex laticincta* (right) that masquerades as it. The moth can be distinguished by its proboscis.

(resembling *Xylocopa* carpenter bees). Apart from sesiids, among the micromoths there are some excellent Hymenoptera mimics, even among the clothes moth family (*Vespitinea gurkharum*), in Cossidae (e.g. the male of *Eulophonotus hyalinipennis*), in Zygaenidae (*Burlacena vacua*), in Thyrididae (*Cicinnocnemis magnifica*) and in Crambidae (*Trichaea pilicornis*). It is noteworthy that a few day-flying micromoth species, e.g. members of Schreckensteiniidae and Scythrididae, raise their long-spurred legs (the hindleg in *Eretmocera laetissima*), in a threatening, bumblebee-like fashion, although they are not themselves particularly like Hymenoptera. Likewise, some hawkmoths make pseudo-stinging motions with their abdomens.

As well as imitating wasps and bees, moths also mimic ants, which typically repel predators using formic acid. The larvae of the Indo-Australian erebid *Homodes bracteigutta* are particularly intriguing as they appear to mimic two weaver ants attached end to end. This resemblance seems to deter predators like birds but may also disguise the caterpillar during its encounters with ants. Fierce workers of the Green Tree Ant (*Oecophylla smaragdina*) seem to ignore these caterpillars (see p. 67). One of the most realistic ant-mimicking moths is *Xestocasis iostrota* (Cosmopterigidae) which mimics an ant head by joining its enlarged labial palps and the ant's abdomen using the enlarged wingtips. Indeed, some day-flying micromoths imitate ants not just in appearance but also behaviour. One unidentified twirler moth species in Madagascar walks backwards on a leaf so that its burnished wingtips mimic the head of an ant, just as in a species of *Xestocasis* that has been filmed in the Philippines. There is also the odd example of apparent Batesian mimicry of other insect orders. Donald Quicke recently revealed a sesiid moth *Akaisphecia melanopuncta* that mimics poisonous red-and-black bugs from the genus *Dindymus* in Southeast Asia. Sesiid moths are generally assumed to be palatable to predators.

It is understandable that avoiding predation by birds may be the principal selective factor behind the evolution of many larger moth patterns and shapes. However, what predators influence the intricate patterns on day-flying micromoth wings? A recent

RIGHT *Xestocasis iostrota*, an amazing ant mimicking a cosmopterigid moth (Cosmopterigidae) from the Philippines.

LEFT Andy Newman has suggested that the early instars of the Lobster Moth (*Stauropus fagi*) caterpillar mimic and even fob off red ants, *Formica rufa*, using their spindly thoracic legs in some kind of tactile communication (the caterpillars have an eversible gland underneath the head which contains chemicals such as formic and acetic acid, possibly assisting in the interaction).

BELOW LEFT *Siamusotima aranea* (Crambidae: Musotiminae) from Yunnan, China brilliantly mimics on its wings the eight legs of a spider. These moths are supposed to mimic crab spiders (Thomisidae), but whether this protects them against spiders or other predators needs to be tested.

discovery has been the remarkable mimicry of jumping spiders (Salticidae) by several unrelated species of micromoths. It was proved experimentally by Jadranka Rota in Costa Rica that four species of metalmark moths of the genus *Brenthia* were able to imitate the territorial displays of jumping spiders as a trick to prevent these predators, strongly relying on sight for prey detection (with their multiple eyes), from jumping on them. It has since been realized that many different moths trick jumping spiders' visual systems through patterns on their wings that do not only mimic the multiple eyes but also the folded legs of these spiders. For example, a large number of moths in the crambid subfamily Acentropinae, whose larvae are aquatic, seem to have this pattern. In New Caledonia there has been a dramatic evolutionary radiation of Micropterigidae, the most primitive family of moths still living. Within this radiation of the genus *Sabatinca*, practically the same jumping spider pattern has evolved at least three times, complete with eyespots and folded legs! In some species the pattern must work when the moth displays it sideways, so that it can mimic the spider legs. In some *Brenthia*, the folded-leg pattern also appears on the fluffed up hindwings. A very similar pattern and strategy is displayed by an unidentified glyphipterigid moth in Hong Kong and a species of *Schiffermuelleria* (Oecophoridae) from Japan. Species of *Siamusotima* (Crambidae) are also particularly striking in the way that dark lines on the white wings mimic spider legs – resembling the posture of crab spiders. The number of mimicked legs in a yet unnamed species of this genus is even anatomically correct. Aside from the case of *Brenthia*, efficacy of the spider patterns on moth wings have not been tested experimentally.

The second type of true mimicry is named after Johann Friedrich T. Müller, a contemporary of Bates, who stressed the advantage that poisonous or distasteful species have, by sharing the same colour patterns, in jointly reducing the cost of 'training' the predator populations. Furthermore, via their common signalling, Müllerian co-mimics more easily cover phenological time (that is the time distribution pattern of different life stages) and fill ecological space with a pattern reinforcing avoidance behaviour by predators. Müllerian mimicry is particularly common within cyanide-protected burnet moths, not just between pairs of closely resembling species like *Zygaena lonicerae* and *Z. trifolii*, but also with more species being involved, either of burnets or even other insects. In the latter case, a complex 'Müllerian mimicry ring' is established. It is not uncommon to observe in a mimicry ring a number of cheating Batesian mimics which consort with Müllerian co-mimics and take advantage of their similarity with the widespread model pattern. Such an assemblage can be seen in a striking Neotropical mimicry ring exhibiting a pattern of a bright orange colour extending across the body and most of the upper wings that contrasts with a zebra-

RIGHT A prominent mimicry ring in moths. In the Neotropical region, members of many moth families have roughly the same psychedelic pattern. From top to bottom rows, and from left to right – 1st row: *Pyromorpha radialis, Euclimaciopsis tortricalis* (both Zygaenidae), *Phostria mapetalis* (Crambidae); 2nd row: *Pseudatteria volcanica, Atteria docima* (both Tortricidae), *Eudule amica* (Geometridae); 3rd row: *Ormetica zenzeroides* (Erebidae), *Eudulophasia tortricalis* (Geometridae), *Calodesma maculifrons* (Geometridae); 4th row: *Pseudomennis bipennis* (Geometridae), *Uranophora walkeri* (Erebidae). The last moth (*Nevrina procopia*, Crambidae; 3rd row, right) is not mimetic with the preceding species as it occurs only in the Asian tropics.

striped wing tip. Members of this ring are the probably toxic species *Pyromorpha radialis* (Zygaenidae), *Napata walkeri* (Erebidae: Arctiinae) and *Oricia homalochroa* (Notodontidae: Dioptinae) as well as doubtfully toxic species of *Pseudatteria*, *Idolatteria* and *Tinacrusis* (Tortricidae), *Mapeta xanthomelas* (Pyralidae), *Phostria mapetalis* (Crambidae) and *Pseudomennis bipennis* (Geometridae).

In mimicry, it is fashionable to flaunt a commonly worn dress that signals status. Some mimics can cheat from the capacity of predators by not going to the length and risk of keeping nasty compounds in their body. Sometimes one sex, usually the female, may have a different appearance, dividing its risk between several 'mimicry rings'. A recently studied example is the Wood Tiger (*Arctia plantaginis*) whose colour varies geographically across Europe. In this extraordinary case, males with black markings that have predominantly white or yellow hindwings coexist in some areas, both implicated to resemble a distinct and potentially Müllerian co-mimic. The yellow form, in particular, appears mimetic of a widespread and unpalatable looper moth (*Arichanna melanaria*). Persistence of the white form, which is more prone to predation, may be otherwise explained by linked genes conferring immunity to viral infections.

Mimicry rings often extend across families, as in the previous examples. Indeed, mimicry between families of Lepidoptera is not uncommon. For example, there are some striking cases of mixed mimicry rings between day-flying moths and butterflies. The large tailed moth *Epicopeia polydora* is a remarkable mimic of both sexes of *Atrophaneura* butterflies (in this case Müllerian mimicry), which in turn are the models for one form of the female of *Papilio memnon* (the latter is a Batesian mimic of poisonous swallowtails in the genera *Atrophaneura* or *Pachliopta*). In New Guinea another swallowtail, *Papilio laglaizei*, is supposed to be a Batesian mimic of the uraniid moth *Alcides agathyrsus*. The last moth is known to contain particular 'polyhydroxy' alkaloids, but the swallowtails might themselves be distasteful.

Mimicry rings can also extend to other insect orders, as seen in some ctenuchine erebids and procridine zygaenids that are excellent mimics of distasteful lycid beetles, particularly in the Neotropical region. Sometimes mimics, either Batesian or Müllerian, do not need to precisely match a poisonous species as long as they have sufficient similarity to a generalized alert pattern. The classic black-and-yellow or black-and-red colourations are well recognized by predators as signalling 'danger', and displaying such colours offers the mimic a major advantage in terms of survival. Several black-and-red burnet moth species commonly gather on flowers among other black-and-red insects such as bugs (e.g. *Graphosoma*, *Spilostethus*, *Rhynocoris* and *Cercopis*) and beetles (e.g. *Trichodes*, *Stictoleptura*, *Acmaeodera*, *Ampedus* and ladybirds, *Coccinella*), not to forget the many black-and-red wasps that often join the feeding fray (e.g. *Anoplius* and *Podalonia*). AZ once observed in Spain a mass gathering of *Zygaena fausta* on flowers together with bugs of the family Lygaeidae (*Horvathiolus* sp.). The patterns were so similar that the moths and the bugs were indistinguishable without close inspection.

Well known in Monarch butterflies, automimicry is a case where less chemically protected individuals can cheat within a species. This occurs when the amount of

HOW MOTHS MAINTAIN FIXED PATTERNS OR VARIABILITY

There is still a lot to find out about how moths defend themselves from attack, and how exactly their behaviour, chemical compounds and patterns are remodelled in a myriad of defensive strategies. Defence from enemies has been a driving force behind some of the most amazing instances of adaptation and diversification in moths. Even as a group particularly poor in parental care, some Lepidoptera have developed strategies to maximize offspring survival. A.D. Blest, a leading authority on saturniid biology, noted that distasteful aposematic (i.e. toxic and warningly coloured) species have little or no pattern variation at all, whereas their palatable, cryptically patterned relatives are highly variable. He also recorded that the first are longer-lived, outliving oviposition for some time too, and have scales that are not prone to come off wings, whereas the others are shorter-lived and the wings become rapidly worn as the scales easily dehisce. These differences seem to reflect the tendency for aposematic species to show stable and conspicuous colour patterns and to maximize their encounters with predators. This trains the latter to avoid such forms so as to confer the greatest possible extent of protection onto their progeny. Conversely, the physical wear of the wings adds to palatable species' pre-existing variability to further diversify their appearance, whilst they die shortly after reproduction. This reduces the chance that predators tune in with a particular search image, which would negatively affect the moths' offspring as well.

Variability in external appearance can be partly determined by multiple variants of the same gene (alleles), each leading to a different phenotype, that is the outward appearance of an organism. Due to the recombination of genes through sexual reproduction, and depending on the number of genes and alleles involved and how these interact, some species may appear in an astonishing number of forms. Such rampant variability is seen in the Tufted Button (*Acleris cristana*) the erebids *Achaea lienardi* and *Ercheia cyllaria*, the noctuid *Cryphia raptricula* and the nolid *Plotheia decrescens*. These are candidates for the most variable moth in the world, the Tufted Button, with some 118 named forms, having been described 37 times as different species and the nolid described 22 times in five different genera. These examples show remarkable polymorphism in forewing pattern but not in hindwings, which are always identical ('monomorphic'). However, when features of the phenotype are controlled by many genes, each with a small additive effect, this makes it difficult to distinguish separate forms. Considering too that some genes produce different effects depending on when they are activated and the environmental conditions at play, the outcome is some species appearing in an almost endless continuum of variation. Examples of such continuously varying moth species are the Japanese Oak Silkmoth (*Antheraea yamamai*), Crescent Dart (*Agrotis trux*) and Clouded Drab (*Orthosia incerta*).

ABOVE Seasonal polyphenism in Early Thorn (*Selenia dentaria*). Spring (top) and summer (bottom) forms of the moth.

RIGHT Polymorphism in Tufted Button (*Acleris cristana*). This moth has 118 named forms and was described as a different species 37 times

The influence of environmental conditions in producing different phenotypes is seen in all those moth species that show seasonal polyphenism. This phenomenon is found when individuals of one generation look markedly distinct from those of another in a season with different environmental conditions. We have already seen above how developing during one or another season leads to strikingly different caterpillars in *Nemoria arizonaria* (see p. 50). The same effect may be found between adults of the spring and summer generations in loopers of the genus *Selenia*, e.g. Purple Thorn (*S. tetralunaria*) where even the wing edge can be variably jagged. Unlike the situations in satyrine butterflies, where the adaptive significance of smaller eyespots in resting among leaf litter in the dry season has been clearly demonstrated, there seem to be few such findings in moths. An exception is Daniel Janzen's demonstration in 1984 in a large saturniid moth, *Rothschildia lebeau*. This has a lighter brown dry season morph that better matches low rainfall leaf tones while a more chocolatey morph, responding when still in its pupa to warmer temperatures, matches the darker and mouldier leaf litter substrate of the rainy season.

In general, predation can drive aposematic species towards a standardized pattern, shift mimics toward different models, and occasionally even push the same species into different mimicry rings. The latter has been demonstrated on a geographical basis in the European burnet moth *Zygaena ephialtes*, which in some regions is a Müllerian co-mimic of other black and red burnet moth species, and in others mimics syntomine erebids such as *Syntomis phegea*, which is predominantly black with white and yellow markings.

Predation is known to maintain polymorphism in palatable moths through a mechanism called apostatic selection. Predators, in fact, tend to optimize their hunting effort by actively searching for the patterns of the most abundant and nutritious prey whilst minimizing energy expenditure. Accordingly, a palatable polymorphic species in effect appears to function like more than one species to predators, so they will selectively prey on the most abundant form. However, this selective predation reduces its abundance, so that fewer individuals are able to reproduce and pass on their genes to the next generation. Alternative forms of the same species will therefore increase their frequency in the coming broods until one becomes common enough that predators start to switch their attention to that, and so on for endless cycles. It is likely that the extraordinary polymorphism in the fleshy caterpillars of the Death's-head Hawk-moth

(*Acherontia atropos*), can be explained by apostatic selection. These have two similarly patterned forms, albeit differently coloured (yellow or green), and a third brown form with a weird dark arabesque-like decoration on whitish thoracic segments. However, if other factors intervene, a rare form may indefinitely remain at low frequency in the population. For example, caterpillars of Angle Shades (*Phlogophora meticulosa*), have rare yellow forms that can survive when birds are hunting for more common brown and green caterpillars but they may possibly not override the other forms (except during leaf fall).

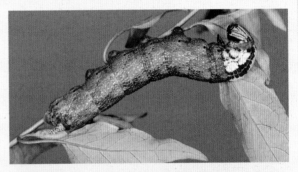

ABOVE Colour morphs of the larva of Death's-head Hawk-moth (*Acherontia atropos*).

toxins or distasteful chemicals greatly varies between kin, e.g. when they feed on plants which themselves show different levels of the noxious substances taken up by the insects. Dishonest automimics presumably occur in a few arctiine erebids, such as the Polka Dot Moth where sometimes caterpillars will feed on non-toxic plants.

An extraordinary kind of resemblance also occurs between the Madagascan uraniid *Chrysiridia rhipheus* and some cetoniine beetles of the genus *Euchroea*, above all *E. clementi*, *E. coelestis* and the appropriately named *E. urania*. Despite these flower beetles having a completely different body shape and size to the moth, their pattern is very similar to that on the moth's wings. The moths and beetles have been observed to visit similar white flowers of forest trees. It is not known yet what kind of mimicry is involved, but the moths are protected by a high concentration of chemicals that are unpalatable to predators. The metallic green-blue scales of the moth are mimicked surprisingly effectively on the beetles against a black background. *Euchroea clementi* even manages to mimic the yellow-gold colour and black spots of the underside of the uraniid moth's hindwing.

Sometimes mimicry rings can cross life stages of the same species, for example in the South African looper moth *Zerenopsis lepida*, larvae and adults are both orange with black spots, and absorb toxins from their cycad hostplants. Well known in the Monarch butterfly, automimicry is a case where less chemically protected individuals can cheat within a species. This occurs when the amount of toxins or distasteful chemicals greatly varies between kin, e.g. when they feed on plants which themselves show different levels of the noxious substances taken up by the insects. Dishonest automimics presumably occur in a few arctiine erebids, such as the Polka Dot Moth where sometimes caterpillars will feed on non-toxic plants.

ACOUSTIC DEFENCES

One of the most important defence mechanisms in moths is the development of hearing. To detect a predator, it really helps to hear it coming. Surprisingly also, many moths 'sing', not only in courtship as we mentioned in Chapter 4 but specifically to defend themselves. There are remarkably few observations of moth songs even from well-worked temperate areas. What would it be like to appreciate forests full of moth song, beyond our range of hearing? Actually some tropical moth songs are well within our hearing range, such as the noise produced by disturbed *Pseudoclanis grandidieri* in Madagascar. However, we are more attuned to noticing colour patterns than subtle sounds, many of which are above the range of human hearing. In museum collections, the rich acoustic world of moths falls silent and can only be imagined by detailed studies of their sound producing organs; this world of acoustic defences, even in the field, is weakly explored.

HEARING Moth ears can detect the ultrasound emitted by bats, prompting moths to close their wings and fall some distance to avoid being snatched by them. About 15% of the 129 moth families, and some 85% of macromoth species, have ears. Except for pyraloids, relatively few micromoths possess ears. These eared moths are known as tympanate, due to the membrane (tympanum) that vibrates at the opening of an acoustic chamber and transmits its motion to a tiny stretch receptor organ. The moth ear can even differ between genders, as in Uraniidae which have their abdominal ears in different places.

Moth ears apparently evolved in the last 50 million years in response to the evolution of bats, but in strictly day-flying moths they must help to avoid attacks by birds and other predators and so be sensitive to lower frequencies. As already mentioned, the gaping holes of moth ears can offer an excellent shelter for parasitic mites. Remarkably, Asher Treat discovered that colonies of the mite *Dicrocheles phalaenodectes* only attack one ear, effectively making the moth deaf in that ear, and if an individual mite is moved, it always finds its way back to the infected ear (possibly following its pheromone trail). This implies that the moth can still defend itself adequately against bats using the good ear, and the mites thus avoid the demise of their host and therefore themselves.

Not all ears have typical tympana. Some hawkmoths (e.g. Striped Hawk-moth) have what is termed a 'palp-pilifer' organ on their mouthparts that is enables them to be especially sensitive to bat attacks, e.g. when they are feeding on flowers. When the palp moves in response to air vibrations, it is the tiny pilifer that detects the high-pitched echolocation soundwaves of bats, transmitting the signal to another receptor inside the moth's head. The palp-pilifer is in all respects a hearing organ, but not a typical ear!

ABOVE The arctiine erebid *Cymbalophora pudica* has particularly enlarged tymbal organs on its thorax.

ALARMS Sound production was discussed in a sexual context (see Chp. 4), but it is often an important part of a moth's defence arsenal. Many erebid moths, above all tiger moths (subfamily Arctiinae) but also tussock moths (Lymantriinae) have sound-producing organs (tymbals) that advertise their distastefulness during flight. AZ once thought he had come across crickets singing, before realizing it was a myriad of Discrete Chaperon (*Cymbalophora pudica*) fluttering around. This moth has possibly the largest tymbals in Europe in relation to body size. Species of *Setina* can also be clearly perceived clacking when they fly, and holding Jersey Tiger moths (*Euplagia quadripunctaria*) close to the ear, a rapid train of clicks can be heard. Some of the moths are highly unpalatable models in acoustic mimicry rings. A good example of an acoustic Batesian mimic is the Orange Beggar Moth (*Eubaphe unicolor*; Geometridae), which while being palatable to bats, mimics the sound of nasty-tasting tiger moths. Just like birds, bats can quickly learn to associate the sounds of tiger moths with toxicity. Sounds are the colours of the bat world and the phenomenon of mimicry in sound (acoustic mimicry) is an expanding field of investigation.

Another way that moths emit sounds is via their wingbeats. The Batesian mimic Poplar Hornet Clearwing (*Sesia apiformis*) closely mimics hornets in pattern and behaviour, and also produces a wing whirr resembling their buzz. H.R. Agee found that when in flight individuals of the Corn Earworm (*Helicoverpa zea*), produce ultrasonic clicks during the upstroke as the tips of forewings touch one another. Simple wingbeating can generate sounds, too, particularly if the strokes are powerful and quick enough such as those of hawkmoths, up to 25 per second in Tobacco Hornworm (*Manduca sexta*), 46 in

RIGHT Orchard Ermine (*Yponomeuta padella*) showing hyaline patches bearing sound producing ridges (inset). Both sexes of ermine moths, even though they cannot hear, mimic sound of other poisonous moths by buckling these tymbal organs. They also advertise toxins in their own bodies via ultrasound easily detected by bats. The phenomenon is known as acoustic aposematism.

Euchloron megaera and 83 in Hummingbird Hawk-moth (*Macroglossum stellatarum*). It is for this reason that Rudolf Mell, following years of experience with hawkmoths in China, considered he was eventually able to recognize which species was fluttering around in his house just by listening to it. A true lepidopterist attuned to moth songs.

The small ermine moths, *Yponomeuta*, offer a nice case of a newly discovered singing mechanism in a supposedly well known genus. In 2017, David Agassiz discovered that the transparent patches at the base of the hindwing of these micromoths bear sets of ridges which differ in number and shape from species to species. Agassiz speculated that these structures make the moths sing, but he was curious that they occur in both sexes. Unbeknownst to Agassiz when he published his paper, in 1996 on the island of Öland, Sweden, Johnny de Jong was recording the ultrasound made by moths and noticed a strange phenomenon in the Apple Ermine (*Yponomeuta malinellus*). Occasionally, one of these nocturnally active moths would lift very slowly from a leaf into the air until it disappeared over the bush-tops while letting out a shrill song that he recorded on his bat detector. What were these ermine moths doing and how exactly do they use their sound organs? Because both sexes have the same structure, and ongoing research confirms they produce ultrasonic clicks in flight by buckling their tymbal organs as tiger moths do, it is likely they are also advertising their distastefulness against bats (a phenomenon known as acoustic aposematism).

Also likely to be a bat defence, but still little studied, is sound production in nolid moths. The same research team discovered another moth that makes an eerie, near 40 kHz ultrasonic song at dusk high in the oak tops, the Scarce Silver Lines (*Pseudoips prasinana*). Its song, first reported by F. Buchanan in 1872, appears cicada-like when it is transformed to lower frequencies, but it is normally inaudible to humans. Actually, the sound had been recorded some 20 years ago, but it was uncertain what produced it. After much work and with a lot of luck, eventually a moth whizzed past the bat detector, and this proved to be the mystery sound source. This species has

LEFT Larva of *Diurnea fagella* showing stridulatory leg (inset). The larva will make a distinct scraping sound with this leg is disturbed or removed from its leaf shelter by a predator.

a pair of tymbal organs on its abdomen. Strangely, it had been known for some time that a noise could be heard out of the boat-like cocoons of silver lines moths, perhaps just before the adult emerges. In *Pseudoips* and several other members of the same family, there is a series of ridges towards the back of the pupa capable of scraping inside the cocoon and making a rustling noise. So remarkably, the pupae of silver lines moths do also sing – using a completely different mechanism.

Even quite primitive leaf-mining moths are capable of using sounds in defence against parasitoid wasps. Candace Low discovered that the larvae of the Tupelo Leafminer (*Antispila nysaefoliella*) have an array of raised warts on the back of the eighth abdominal segment that they use to scrape the top of the leaf mine when parasitoid wasps try to lay eggs through the mine. Similarly, Lynn Fletcher and colleagues discovered that larvae of *Caloptilia serotinella* (Gracillariidae) have a repertoire of plucking, vibrating and scraping noises within their leaf mines. In this case, the noises are produced most frequently when another caterpillar intrudes into the mine. The larvae of the March Tubic (*Diurnea fagella*), appropriately named in German 'Die Sangerin' (The Singer), as well as other Chimabachinae (Lypusidae), exhibit an inflated tibia on their hindleg pair, from the third instar onwards. These 'paddles' are alternately drummed using tarsal spines in contact with the leaf surface, producing a scratching noise audible to humans. This tantrum display occurs when the caterpillar is disturbed out of its silken shelter.

We have seen some of the range of predators and how moths fight back against them in all stages. Next, we address the ecology, distribution and importance of moths, looking at their extraordinary diversity and some of the main underlying ecological factors.

CHAPTER 6

Diversity and distribution

WHY IS THERE SUCH AN EXUBERANCE OF MOTHS? Here we address how many moths there are, where they are richest and where the biggest gaps remain in our knowledge of moth diversity. Then we examine the main ecological factors that can explain this diversity, in particular the moths' interaction with plants. What are the most extreme environments in which moths live? Where did moths originate and how do they get about the planet?

A CORNUCOPIA OF MOTHS

In 2011, a team of 44 lepidopterists networked in an attempt to count all species of moths (as well as butterflies). They documented 138,885 moth species names that are currently regarded as valid, i.e. described species, which belonged to 13,772 genera. Although this sounds very precise it is actually an estimate, because updated catalogues are lacking for many moth families. It also ignores how many species remain undiscovered, especially in the tropics. Some countries appear to be extremely rich in moths, yet still remain black holes in our knowledge. For example, there are no described Gracillariidae leafminers from Bolivia, which surely has hundreds of species in this family. The true species richness of moths is therefore a much larger figure, and our knowledge of their diversity is constantly evolving.

In some of the Earth's remotest places, such as the Hawaiian island chain, most moth lineages common in continental areas are missing. Nevertheless, a single lineage has gone crazy. Daniel Rubinoff and Klaus Sattler estimate that the exclusively Hawaiian genus *Hyposmocoma* (Cosmopterigidae) has as many as 330 species. What is more, most species occur in the geologically recent part of the volcanic island chain, which is 1–5 million years old, implying that their evolution was relatively rapid; however, it is possible that ancestral species are concentrated in the northwest islands, of the much greater age of up to 21 million years. Nevertheless, it would actually be unusual to find so many species in a single genus in a single country from a mainland area of equivalent total surface area (e.g. Albania with 28,750 km²). Many more species can occur in a large continental region, for example, the genus *Coleophora* has more than 1,200 species principally across the entire Holarctic.

OPPOSITE Adults of the cosmopterigid genus *Hyposmocoma* showing the phenomenal diversification in size and pattern undergone by these moths in the Hawaiian islands.

An approximate method to estimate total species numbers is called extrapolation. In the relatively well studied British Isles, 2,217 moth and 59 butterfly species are resident. In the 2011 Lepidoptera count mentioned above, there were 18,539 butterfly species validly described in the world. If we consider that butterflies are far better known than moths and assume that the ratio between butterflies and moths is worldwide the same as that for the British Isles, then we would expect around 700,000 moth species to occur globally. Undescribed moths might therefore outnumber named ones by a factor of about five, although such extrapolative estimates are highly speculative for now. A flaw in such reasoning is that because numbers of species in moth and butterfly families are far from evenly blended between regions, the real number might differ greatly. The sad fact is that inevitably many species are lost every year by forest destruction without us ever knowing. Those highly vulnerable moth species occupying the smallest geographic areas are usually the least known of all.

In order to understand what knowledge currently exists about the diversity of moth species, we must first untangle their names (see pp. 138–139). The name validly applied wins priority over other subsequent names given to the same species. Far from all species that have been named belong to moth species that are today regarded as valid species. The names that are not valid are known as synonyms, and in some large groups, a quarter or a third of names might not be valid. This is an important reason why most catalogues of moths are estimates.

Moth catalogues though can be extremely valuable. Given the many decades if not centuries of work making sense of existing taxonomy, some crude worldwide patterns can be discerned by carefully building lists of species per country or region. The online database Afromoths, for example, attempts this task for the entire Afrotropical region. Adding to or even independent of this taxonomic knowledge, the number of species in a sample can be worked out locally, either by examining and genitalic dissection of all individuals, or by using a direct local surrogate for their richness (see DNA barcoding, p. 140). We know from these approaches that the Neotropical region, particularly South America, is particularly rich in moth biodiversity. We can be sure also that species richness (the counted number of species in a sample) tends to increase with land area, elevation and environmental diversity. Certain families such as the Saturniidae are extraordinarily rich in the New World tropics. Other families are exceptionally rich in particular regions and environments. Moths of the family Oecophoridae that feed as larvae on dead leaves (these are so called leaf litter detritivores) are richest in Australia, where there are at least 2,300 species, a tenth of its still partially catalogued moth fauna.

Generally speaking, the richest and most diverse regions for moths are tropical ones. Some moth groups, usually at the level of large genera, pose exceptions to this rule, showing far more species in temperate zones than in the tropics. Examples include *Phyllonorycter* (Gracillariidae) – around 340 species across the Holarctic region – *Coleophora* mentioned above for the same region, and the noctuid genera *Catocala* (Erebidae) with over 225 species and *Zygaena* with over 100 species.

The rich diversity of moths in the tropics is due to numerous factors, most important probably being the number of extra generations possible in rainforests due to the lack of either a winter or pronounced dry season. With extra generations, there is more chance for mutations and evolutionary diversification. Combine with four other important factors: the vast areas covered by lush tropical forests, especially in the past, the structural complexity of rainforests with differing habitats from the forest floor to the canopy, the far greater density of plant species than in arid or temperate zones, and the longer term climatic stability of the tropics relative to regions which were severely glaciated at periods during the Pleistocene, and it is apparent that the potential for tropical forest speciation (see Chp. 7) is tremendous. The Equatorial belt is among the richest places on Earth for moths – in the Neotropics especially, where it intersects the Andes. Species ranges overlap more frequently where the lowlands meet the mountains. The general rise in species richness as we move towards the Equator is a phenomenon that depends not just on latitude, but on the area of land that was available in the past for evolution and species radiation.

BELOW Parade of *Catocala* moths illustrating the diversity of colour pattern in a single genus.

THE DELIGHT OF NAMING MOTHS

TAXONOMY, SYSTEMATICS AND NOMENCLATURE
Taxonomy is the theory and practice of classifying organisms. Taxonomy links directly to the discipline of Systematics, whose practitioners attempt to uncover the patterns of evolutionary relatedness between organisms. By combining the results from Systematics with a set of procedures and rules, taxonomists establish hierarchical systems of classification which provide the current framework for the 'Tree of Life'. Zoological Nomenclature underpins Taxonomy via the naming of groups which are thus identified.

ETYMOLOGY OF COMMON WORDS
Where do the words caterpillar and moth come from? The earliest mention of caterpillar is in 1440 when the French referred to 'catyrpel', meaning 'hairy cat' (a Latin derivation of 'catta pilosa', and later the French use of 'chatepilose'

ABOVE Angle Shades (*Phlogophora meticulosa*) is one of the more familiar and aesthetically named noctuid moths.

meaning 'shaggy cat'). The word pupa means 'young girl, doll' (in the sense of pupil meaning undeveloped). The word moth by contrast has a north European origin. It comes from the old Norse 'motti' (middle Dutch and German 'mot' or 'motte' and old English 'moooe'). Up to the 1500s the word referred mainly to the larvae of clothes moths. There might also be an onomatopoeic element as in the Italian 'farfalle' - recall the effect of passing moth brushing your face at night!

COMMON NAMES
Several countries enjoy common (vernacular) names for their moths and some of these names are steeped in history. The Blue Underwing (*Catocala fraxini*) is otherwise known as the Clifden Nonpareil. It was first found in England in or before 1749 – not at Clifden near Bristol, but at the Cliveden 'Cleifden' Estate along the River Thames in Buckinghamshire. Mr Davenport spotted it 'sticking against the body of an ash tree', according to a work by Benjamin Wilkes. Wilkes' name 'nonpareil' means 'without equal'. Such evocative English vernacular moth names mostly date back a century or more, and still never fail to appeal. They come in many categories: romantic (True Lover's Knot, Maiden's Blush, Scarce Brindled Beauty), human behaviour (Old Lady), identification uncertainty (Suspected, Uncertain, Nonconformist), bestowing nautical or aviation allusions (Jersey Tiger, Gypsy Moth), based on people (Blomer's Rivulet, Barrett's Marbled Coronet), places (Rannoch Sprawler, Kentish Glory), caterpillar habits (Drinker), on patterns (Mother Shipton, Shuttle-shaped Dart, Setaceous Hebrew Character, Angle Shades, Lunar Spotted Pinion, Scarce Merveille du Jour, Brown Line Bright Eye, Buttoned Snout), or colours (Burnished Brass, Green Brindled Crescent, Crimson Speckled Footman). Geoffroy's Tubic is a rare example of an English micromoth with an uncontrived and elegant name.

SCIENTIFIC NAMES
The scientific names of moths follow the time-honoured tradition of binomial nomenclature (genus and species), as decreed in *Systema Naturae* (1758) by Carolus Linnaeus. The names, primarily Latin (sometimes Greek-derived), are crucial for scientific communication, as only the Latin name is universal between languages. Early genera were vast in their scope, sometimes as broad or broader than today's families (*Tinea* for micromoths, *Noctua* for night moths, *Geometra* for looper moths, *Phalaena* as a general name for moths) while species names (e.g. respectively, *pellionella*, *pronuba*, *papilionaria*, *typica*) were often as narrow as they are today.

WHY ARE MOTH NAMES BASED ON ADULT SPECIMENS?

It is a curiosity of entomology, not least the study of moths, that an insect's name was almost always based on a single preserved specimen (the holotype) in the adult stage. There is nothing in that taxonomic bible, the *International Code of Zoological Nomenclature* (ICZN), forbidding naming based instead on early stages. Yet we have no examples of Latin names founded on 'immature' moth holotypes, despite their preservability, or even 'pinnability' (although whether blow dried or in alcohol, they do not so easily keep their shape and colours). In the past this made sense, because someone finding an adult moth would not always reliably be able to associate it with a name based on a caterpillar. Today this needs no longer apply, since through DNA, it is possible to link immatures with adults (see p. 140, DNA barcoding); and some species may only ever be collected in an early stage. However, conventions run deep in taxonomy and particularly among lepidopterists, who prefer to stick with a nomenclatural prism of the natural world viewed through the adult form.

DEDICATIONS

Can taxonomists describe moths to honour themselves? Yes and no. If a moth's name is dedicated to a person, the convention is to name it after someone else other than the describer. There is at least one moth example that flouts this rule: Sigismund Hochenwarth in 1785 named a species as *Phalaena hochenwarthi* (now *Syngrapha hochenwarthi*, a handsome moth with yellow hindwings). Apparently Hochenwarth got an undeservedly bad reputation about this because the suggestion for the name had been given to him. Again, there is nothing in the ICZN forbidding this but it is a matter of scientific and professional humility not to name a moth after yourself.

ABOVE Clifden Nonpareil (*Catocala fraxini*; Erebidae).

Offensive names are not permitted, but there can be humour in the naming. Some moth names reveal the obsessions of whoever named them. In 1964, prolific grass moth taxonomist Stanislaw Bleszyński named a moth from the genus *Pseudocatharylla*, '*gioconda*' – bringing to mind the 'Mona Lisa' – because it was 'very distinctive' and was described from 'a unique female'. After two divorces though, he steered clear of naming moths after ladies. In 1966, he named one of the shortest moth genera '*La*', including in it *La cucaracha* and *La paloma*. Bernard Landry in 1995 finally found a third species and named it, refreshingly, *La cerveza*.

When taxonomists want to name a large group of species they are sometimes imaginative, unlike William Kearfott, for example, who named a large series of tortricid moths *Eucosma bocana*, *E. cocana*, and so on *ad nauseam* up the alphabet. By contrast, the late Gaden Robinson in his 2008 monograph of scuttling moths, *Edosa*, introduced a fine set of dedications to rock and jazz musicians including *E. bodiddleyella*, *E. claptonella*, *E. hendrixella*, *E. leadbellyella*, *E. longhairella*, *E. mayallella*, and to cap it off, *E. screaminjayella*.

Sometimes a dedication can become immortalized at a higher taxonomic level than the original. Paul Whalley, a tropical leaf moth specialist, had a genus from Madagascar named after him in the family he studied, Thyrididae, by Pierre Viette in 1977. Within this genus, *Whalleyana*, Viette included two species, *W. toni* and *W. vroni*, placed appropriately in Whalley's specialist family. In 1991, Joël Minet realized that the moths justified a new family, Whalleyanidae. They did not fit in the superfamily Thyridoidea nor elsewhere so, by default, these Malagasy moths then became members of a new one, Whalleyanoidea.

Indeed, innumerable moths have been named after humans. The converse is rare, however: few people have been named after moths. Hannah Rothschild, in *The Baroness*, described how her great-aunt, Nica de Koenigswarter, got her Christian name Pannonica from *Eublemma pannonica*. Yet in her shadowy jazz life, Nica wanted to be known as a butterfly, and in fact, the song *My Little Butterfly* by her idol Thelonius Monk was dedicated to her. Hannah was shocked when she visited the Natural History Museum, London and was shown, instead of a butterfly, the small moths that Hannah's grandfather Charles had collected around 1910 in Hungary. However, Hannah remarked that Miriam Rothschild's disdain, when the subject of her sister Nica came up, was appropriate because she was a 'creature of the night' and 'only came alive after dark'.

DNA BARCODING

A section which is 658-nucleotide base pairs long of the mitochondrial gene Cytochrome Oxidase I (COI) has been adopted as a standard 'barcode' to tag species across the Animal Kingdom (the DNA barcode). This DNA fragment is long enough to differ substantially between almost all animal species. Once samples (for moths usually just a single leg works well) have had their DNA extracted and sequenced, resulting COI sequences can be compared between individuals of a single species and with other species. Trees of DNA similarity (approximate relationships) can be built based on these sequences. The quickest procedure used by BOLD (Barcode of Life Database), is a clustering program that calculates the percentage differences in nucleotide identity between and among compared sequences. These differences are expressed as an array (known as the distance matrix) and then a computer program rapidly builds a dendrogram (known as a neighbour-joining tree) that optimizes the relationships between the sequences based on the simple distance matrix. In cases where several individuals of a species have been sequenced, the branches of such trees comprise groups of putative species (the majority of species are at least 2–3% different from members of the nearest 'species' cluster). By contrast, differences within a species are usually an order of magnitude less, with different individuals from across a species' range typically differing by only fractions of 1%.

The 'DNA barcoding' system has been extensively applied and works well for maybe 95% of species that had previously been distinguished on a morphological basis. It was first tested in 2003 by DNA barcoding founder Paul Hebert when he investigated the contents of a moth trap in his back yard in Canada. Results were good enough that practically every local moth species could reliably be distinguished. In recent years, the system has been further developed, by allocating a unique code to each well differentiated cluster of sequences using a sophisticated computer algorithm. For example, BOLD:AAA0001 is the DNA Barcode Index Number (BIN) for *Homo sapiens*. If you type the number into the online available Barcode Index Database you will notice that this BIN includes all barcoded human populations around the planet and there is even a single barcode for a fossil human '*H. heidelbergensis*'. The BIN system is effective because it provides a sort of surrogate taxonomy. This is incredibly useful for ecologists for estimating species richness, and the species turnover from site to site, as you can compare the composition of BINs as proxies of real species between sample sites. This works whether the sequenced organisms already have a zoological name or not. Even as yet undescribed species can thus be included into ecological studies without examination by taxonomists.

The BOLD database contains a huge number of sequences – for Lepidoptera alone, currently, over a million DNA barcodes for more than 113,000 species that are already described or need to be. DNA barcoding is therefore a powerful identification tool, as long as sufficiently related sequences to the one being queried are present. In many instances it also allows 'mini barcodes', i.e. COI fragments of say 150 base pairs, to be identified successfully. This means that old and degraded moth specimens, or small fragments of DNA in e.g. bat dung, can be identified, even allowing analysis of extinct species or predator diets. It is even impacting correct labelling of food products. A simple way to start to use the DNA barcode database is to type a family, genus or species name in the taxonomy tab and see what DNA barcodes have already been recorded for that name.

BELOW The trace represents an accurate readout of the DNA barcode of *Acronicta retardata*. Each of the 658 bases in the DNA sequence is represented by a colour code representing Adenine (A, green) Guanine (G, black), Cytosine (C, blue) and Thymine (T, red).

0   243

244   487

488   631

INTERACTIONS WITH PLANTS

We have discussed how many moth species are known and might exist. Now we look at the main ecological relationships and unusual environmental niches hosting this richness. The most important ecological relationship of moths is that between their larvae and the hostplants on which they feed. There are over 370,000 flowering plants and the number of moths would probably turn out to be greater (see p. 136), if or when tropical moth inventories are complete. Add to that 1,000 gymnosperms, 11,000 ferns and horsetails, and 20,000 bryophytes including mosses, liverworts and hornworts, in addition to other photosynthesizing organisms (15,000 lichens as well as algae). We detail some of those food choices below. We talked about fungi in Chapter 3 (p. 64). There is definitely not a 1:1 relationship between diversification of moths and plant species. Indeed, it is common to find plants that are not fed on by moths. In cases where moths are specialized on particular groups of plants, let us say a very diverse plant genus, the phylogenetic pattern (tree of evolutionary relationships) inferred for the moth group does not usually match that of the plant group. What seems to happen, in studies that have been done, is rather that moths exploit the existing plant diversity in a very biased fashion, often being very choosy in their tastes even when they speciate, but sometimes shifting in their own evolution to unrelated species. In other words, plants and moths do not evolve neatly in lock step, even though plant diversity is fundamental to moth diversity, as they are a main set of herbivores.

LEFT Adult female asket-cocoon parasitoid wasp (*Meteorus pulchricornis*) threatening her host, a caterpillar of the Corn Earworm (*Helicoverpa armigera*).

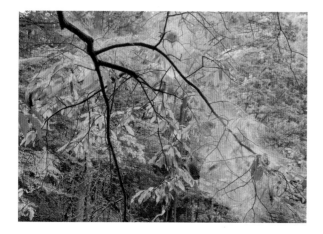

TOP Webs of the Fall Webworm (*Hyphantria cunea*). This is one of the world's most polyphagous moths, of North American origin, but recently introduced into Europe and Asia. It is known to feed on at least 636 plant species, including almost any deciduous tree, and many conifers and herbs.

ABOVE Larvae of Painted Schinia Moth (*Schinia volupia*) feed only on the genus Gaillardia and the adults are also spectacularly adapted to match blossoms of their host plants, here on Indian Blanket (*G. pulchella*).

BROAD AND NARROW DIETS

An important factor in diversity as well as evolutionary success is whether a moth has a broad or narrow dietary range. A single species, or a much larger lineage, may either have a relatively narrow (specialized) or large (generalized) diet breadth. Polyphagous moths (which can feed on a broad range of plants) tend to have sophisticated gut detoxifying mechanisms that allow them to overcome the chemical defences of a wide variety of plants. This can thus enable a moth to exploit many plant species of different families and even become a pest of crops. An example of such a pest is the Mediterranean Brocade (*Spodoptera littoralis*). Because of their ability to feed on numerous kinds of plants, such polyphagous moths can be more abundant and have wider geographic ranges. Specialists, though, rely on plants from one genus or genera of a family (so called oligophagous species) or just one set of closely related plant species (monophagous). Such moths can only become widespread not only if the hostplant has a wide range but if they are also dispersive. Often though, such monophagous moth species have narrow geographic ranges, much smaller even than the plant they feed on. They may often be restricted to feeding on just one part of a plant (e.g. moths that feed on ripening seeds).

If ecologists could only know all the hostplant species in a particular habitat (e.g. a tropical forest), the number of moth species per host, and what proportion of them are specialized in their hostplant requirements, they could extrapolate how many species might really occur there. Vojtech Novotny and colleagues, in long-term studies of tropical herbivores, particularly moths, thus confirmed that tropical herbivores are on average more specialized than ones occurring in temperate zones. These specializations are important factors, in conjunction with others (p. 137), explaining the high diversity of tropical moths.

Moths interact with plants in many and complicated ways that go beyond generalist or specialist association with their host plants. Some moths with vast numbers of eggs are like Lancaster bombers, spreading their eggs far and wide. It is not surprising that such careless mothers are either of polyphagous species or feed on extremely abundant plants, such as roots of grasses in open grasslands (e.g. swift moths on the Tibetan plateau). Moths with females with rudimentary wings, like the Winter Moth (*Operophtera brumata*, p. 31), also tend to be polyphagous, because the larvae, which disperse through the air on strands of silk (a behaviour called 'ballooning'), must make do with the available shrubs and trees wherever they land. By contrast female moths that have very choosy larvae must seek cues as to which plant to lay their eggs on. These cues may be visual, tactile or more often chemical. However, their main enemies, which are the parasitoid wasps and flies (see p. 99), use the same specific plant volatiles (known as kairomones) to zoom into the right place where their moth hosts can be found.

CONSTRAINTS IN COLONIZING PLANTS There are some important factors and constraints to consider in moths colonising new plants. Not only is there a staggering variety of plant species to dine on, the moth-plant interface is an ever-shifting dynamic in time and space, a chemical and ecological battleground that engenders a rich range of adaptations resulting in breathtaking diversity. Random events, such as a moth ovipositing on a plant beyond its normal hostplant spectrum, can lead to the colonization of a whole new plant family. Such shifts are rare but a successful one may well be accompanied by the coincidence of a preadaptation. For example, in zygaenid moths the ability to synthesize toxic cyanide-releasing compounds as a defence against predators was coupled with the moths' existing adaptation to be resistant to them, paving the way to colonizing plants like the cyanide-laden Bird's Foot Trefoil (*Lotus corniculatus*). The diversity of hostplant associations among moths today is testament to the many successful shifts that happened in the past.

Apart from preadaptations favouring host plant shifts, one of the influential ideas in hostplant ecology is that of 'apparency'. Plants that offer a vast resource in both space and time and even in the number of micro-habitats that they provide would be, according to this theory, more 'apparent' to herbivores that might interact and then successfully colonize them. Such plants might host more moth species. A classic example is the English Oak (*Quercus robur*), which has both great 'architectural' complexity and covered vast swathes of Britain until recent times. Accordingly, Richard Southwood tallied 187 moth species feeding on English or on Sessile Oak (*Quercus petraea*) whereas just 40 on Beech (*Fagus sylvatica*). A generally positive relationship between moth richness and surrogates of apparency such as relative abundance, tree height and long term presence for 25 genera of trees has been confirmed by Martin Brändle and Roland Brandl for Germany, where *Quercus* also ranked the highest with 305 Lepidoptera herbivores compared with 119 species on *Fagus*.

LEFT Larva of a Five-spot Burnet (*Zygaena trifolii*) feeding on Bird's-foot Trefoil (*Lotus corniculatus*). The ability to synthesize cyanide-containing compounds means that the caterpillars are resistant to cyanogenic plants such as this legume.

GYMNOSPERM AND FERN FEEDERS Not surprisingly, most moths feed as caterpillars on flowering plants (angiosperms), while moths that feed on gymnosperms are far fewer. For example, among the 50,000-odd named moth species in the Americas, 787 species from 13 superfamilies are so far known from conifers, 291 of which generalists, most of those being Macroheterocera (p. 4) that also, however, feed on angiosperms. Although the relative number of angiosperm feeders overall has not been tallied, it would appear that gymnosperm feeding is much more prevalent than expected from the very low number of gymnosperm species globally compared to angiosperms (only c.1,000 vs 370,000). A degree of polyphagy among moths can only partly help explain this. A more important factor is again apparency: vast areas (particularly northern and montane habitats rather than lowland tropics) are covered by conifer forests, and gymnosperms have been around for a longer geological time. Members of many moth families, including Agathiphagidae, Aenigmatineidae, Argyresthiidae, Tortricidae, Pyralidae, Geometridae, Lasiocampidae, Sphingidae, Saturniidae and Noctuidae, have gone in for feeding on pines, cypress and other gymnosperms, with most micromoth families relatively specialized on them. Such moths have developed means to deal with the mechanical and chemical effects of the terpenoids and other chemicals contained in resins of these plants. The genus *Milionia* – comprising looper moths that happily defoliate hoop pine trees (*Araucaria*) – has radiated into a plethora of exquisitely metallic species involved in mimicry rings in the Papuan region. Some species of *Milionia* sequester noxious chemicals other than terpenoids, such as lactones and glucosides, which make them distasteful to predators (see p. 108). Some of the oldest gymnosperms are cycads; South African cycads have been colonized by the gorgeous looper moths in the genus *Callioratis*.

Even with up to 11,000 fern species (pteridophytes), fern feeding is restricted to very few groups of moths, and there is a lower ratio of fern feeders among moths than ferns among the plants. This lines up to some extent with a lower apparency and architectural complexity in these plants, but fern feeding is also a

BELOW LEFT The Podocarpus-eating larvae of *Milionia basalis pryeri* in Taiwan. Feeding on a gymnosperm is considered to be quite an unusual habit in moths, as the caterpillars must deal with resins.

BELOW RIGHT The fern-eating larva of *Callopistria floridensis*, feeding on Chinese Ladder Brake (*Pteris vittata*). Fern-feeding moths belong to a few specialized groups such as those of the genus *Callopistria* among noctuids, whose larvae resemble parts of a fern frond.

very specialized lifestyle that arose rather few times. A few moths feed on their spores, such as adults of the primitive family Micropterigidae and larvae of a few tineids like the Hart's-tongue Smut (*Psychoides verhuella*). A few hepialid larvae feed on the rhizomes (underground stems) of ferns like Bracken, while most species feed on fronds. These include caterpillars of lithinine geometrids, e.g. Brown Silver-line (*Petrophora chlorosata*), a few crambids in the subfamily Musotiminae and the genus *Hoploscopa*, and those of eriopine noctuids. The latter are beautifully frond-like in colour pattern, the case for the Latin (*Callopistria juventina*). Larvae of the Old World butterfly moths (Callidulidae) in the Southeast Asian tropics feed exclusively on ferns. Females lay their flattened, often translucent eggs, which look like scale insects, on fresh fronds.

How moths such as callidulids came to specialize on ferns is unknown. This is intriguing because fern-feeding larvae must be able to cope with an unusual arsenal of plant defences. Bracken (*Pteridium aquilinum*) for example, contains ecdysones, chemicals associated with moulting activity. They can induce caterpillars literally to moult to death. Despite its remarkable 'apparency' due to the vast land areas covered by stands of this species worldwide, relatively few moth species actually feed on Bracken. John Lawton spent part of his career looking at this – in Britain, he documented only seven moth species that regularly use it, actually about the expected number given Bracken's low architectural complexity.

ALGAE AND LICHEN FEEDERS Although the vast majority of moths associate with flowering plants rather than gymnosperms or ferns, there are groups adapted to feeding on other lifeforms that photosynthesize. In clean air and mountain areas, lichens are important for the few moth groups that specialize on them. In particular, erebid moths in the tribe Lithosiini (many of which are known as footman moths) feed as larvae either on algae, or specialize on lichens, probably driven by the algal component of these algal-fungal sandwiches that encrust twigs and rocks in unpolluted environments. This specialization has enabled them to radiate into astounding diversity worldwide. Some noctuid moths (subfamily Bryophilinae), including the Tree-lichen Beauty (*Cryphia algae*), are also associated with lichens in addition to mosses.

ABOVE The lichen-eating larva of the Brussels Lace (*Cleorodes lichenaria*), blending in perfectly to its lichen food source (here *Hypogymnia physodes*). Lichens are a really unusual food choice for a looper moth.

BELOW LEFT The leafy liverwort-eating larva of *Sabatinca aurella* in New Zealand, also matching its liverwort food source, usually *Heteroscyphus normalis*. Liverworts are very primitive land plants, and micropterigids are the most primitive of extant moths.

BELOW RIGHT Moths among mosses. Caterpillars of the pyralid subfamily Scopariinae build silken tunnels among grasses and moss. Many scopariines eat moss, as here for the Little Grey (*Eudonia lacustrata*), otherwise an unusual diet for moths.

RIGHT Galls on a Pepper Tree (*Schinus longifolius*) made by the primitive cecidosid moths *Cecidoses eremita* (left) and *Eucecidoses minutanus* (right). Their exit holes resonate in the wind giving the plant its local name meaning 'Whistler'.

RIGHT Galls on a Pepper Tree (*Schinus longifolius*) made by the primitive cecidosid moths *Cecidoses eremita* (left) and *Eucecidoses minutanus* (right). Their exit holes resonate in the wind giving the plant its local name meaning 'Whistler'.

BRYOPHYTE FEEDERS (MOSSES AND LIVERWORTS) The moths that feed on mosses and liverworts are very special indeed. The pyralid subfamily Scopariinae go in for feeding on mosses, among which their caterpillars make silken tunnels. Some micropterigid larvae, such as those of the genus *Sabatinca* in New Zealand, specialize on different species of liverworts. In Japan though, Yume Imada and colleagues discovered that no less than 25 micropterigid species feed on the same liverwort species (*Conocephalum conicum*), but with each species distributed in a different location of the Japanese islands. Liverworts are some of the most primitive plants, but in Japan, this discovery shows that plant diversity does not in all cases beget moth diversity. In that case it appears moths have diversified merely by evolving over long periods of time in different places (see allopatric speciation, p. 163).

CATERPILLARS THAT INDUCE PLANTS TO PROTECT THEM One more extreme interaction of moths with plants that affords them protection against climatic extremes in their environments (be they dry or moist), is that of the galling lifestyle. Moths associated with galls include members from several families (e.g. Cecidosidae, Gelechiidae, Gracillariidae, Momphidae, Sesiidae, Thyrididae, Noctuidae). Larvae of these moths have the capacity to induce galls in plants, which give them hardened protection against climatic vicissitudes, predators and parasitizing wasps. The galls also allow the caterpillars to divert or control plant resources as necessary for their feeding and survival.

Caterpillars of Goldenrod Gall Moth (*Epiblema scudderiana*) build up glycerol as an antifreeze in their tissues as a result of their metabolism during the autumn in quantities up to a fifth of their weight, which enables them to endure temperatures down to -38°C (-36.4°F). *Cecidoses eremita* is one of the many gall makers on the Chilean Pepper Tree (*Schinus polygama*). Galling cecidosid moths in Brazil are specialized for dry environments. The moths emerge from a round flap in the gall. This flap whistles in the wind, hence the local name for the plant '*assoviadeira*' (whistler).

Gall making Lepidoptera are not at all well studied in the tropics. Even one recently described neotropical pigmy moth, *Stigmella gallicola,* makes a gall on the tree *Hampea appendiculata* in Costa Rica. Only a few gall makers in family Nepticulidae were previously known, including the Poplar Petiole Gall Moth (*Ectoedemia populella*). Larvae of the nolid *Garella nilotica* have also been frequently found in galls on tamarisks, though it is unknown whether they actually eat the plants or gall-inducing mites.

PROBING MOTH-PLANT ASSOCIATIONS

The full range of ecological associations of moths can only be guessed at. Modern technologies have the potential to probe the vast museum collections of adult moths for the presence of chemical compounds or isotope ratios that reflect what they fed on as larvae or what plants they utilized to feed on as adults. For example, analysis of the Carbon^{13}/Carbon^{12} isotope ratio in insect parts can reveal the specific photosynthetic pathway (the two main types are named C3 or C4) of the plant(s) that they fed on as larvae, as plants of the two types differ reliably in such a ratio. This method works particularly well for grass feeders such as many Noctuidae. Larvae can be collected in the wild and the hostplant remains in the mouth or gut can also be DNA barcoded. It is thought that DNA, however, does not survive metamorphosis to reveal the plant that a caterpillar ate, although this idea has not been widely tested. It is already feasible, though, in some instances to recover DNA from an early stage revealing the parasitoid wasp species that killed it; this works rather well for parasitoids that emerge from pupae. There are enormous possibilities in the future using molecular or physical technology to probe the deeply entangled food webs of moths. DNA sequencing is so powerful today that it can be used to probe the ecosystem that once surrounded a moth specimen, e.g. via pollen grains and spores attached to it or to assess the identity of inadvertently pressed caterpillars in leaf mines among herbarium sheets.

HABITATS

Moths live in a wide range of environments. Where rainfall is high (as in the tropics) and where different habitats and elevational zones overlap or abut closely (for example, when lowland forests are adjacent to large mountain ranges), the number of species tends to be high, for a wide range of factors. Yet moths can be found in habitats of very low rainfall and some moths are specially adapted to dryness. Extremophiles are lifeforms occurring in environments that would be considered very hostile. We mention below some of the instances where moths have adapted to living in freshwater, hot dry environments like deserts and the extremes of cold such as at high latitudes and elevations (defined as vertical distances above sea level, as opposed to altitudes above the ground), and areas of high ultraviolet radiation or even in total darkness.

DESERT MOTHS

Moths can be found in surprisingly arid habitats and some are specially adapted to desert environments, which challenge moth physiology. Many larvae of desert moths

survive and resist desiccation by feeding inside the hostplant or its fruit (for example, *Endoxyla*; Cossidae) and spending more than a year as a larva – the adult stage by contrast is very short and and makes do with its larval food reserves. Early stages may also proceed extremely fast: from egg to pupa in 19 days for many Heliothinae noctuids. Roman V. Yakovlev has found a rich diversity of carpenter moths (Cossidae) that range through the major deserts of the Palaearctic. The symbiotic relationships of yucca moths with yuccas such as the Joshua tree (*Yucca brevifolia*) have already been covered (p. 71). James Tuttle mentioned that the Phaeton Primrose Sphinx Moth (*Euproserpinus phaeton*) rather than hanging its pumped up wings downwards under gravity to dry, a process that takes about half an hour, rapidly expands them upwards just after dawn in the Mojave Desert before the desert air becomes too dry. Pupal diapause in such environments can be very long; Jerry Powell recorded prodoxid pupae to hold over as much as 30 years!

AQUATIC MOTHS

Moths are not associated with marine environments at all, although some can fly huge distances across the sea as discussed below. Freshwater environments, though, are some of the most challenging ones for moths to colonize, yet some have done so. Probably less than 0.5% of moth species are truly aquatic. A few caterpillars can actually be semi-amphibious. For example, some Micropterigidae caterpillars live in such damp places that they can survive partial immersion in water film. The larvae of *Epimartyria auricrinella* have been inferred to go a step further and make use of a plastron gill (a device to trap air bubbles) so they can be completely submerged for intervals. By mining reeds, the pyralid subfamily Schoenobiinae has devised another method to colonize lakesides and freshwater to avoid submersion, in effect using reeds as snorkels, as do many wainscots (Noctuidae).

With over 750 species worldwide, the crambid subfamily Acentropinae is by far the most significant moth radiation into freshwater environments. Larvae of some aquatic

LEFT The caterpillar of *Eoophyla* sp. (Crambidae) exhibits tracheal gills as an adaptation to an aquatic environment. Other aquatic caterpillars rely on air bubbles, so called plastron gills, for breathing underwater.

acentropines like chinamark moths of the genera *Parapoynx* and *Eoophyla* bear feathery gills from their spiracles that allow them to breathe underwater and then feed on plants like water ferns (*Azolla*). As mentioned in Chapter 1 (p. 34), the adult females of Water Veneer (*Acentria ephemerella*) have fringed hairs on their legs that allow them to dive under water and lay eggs in water plants. Surprisingly, the females have two forms. Most of these have rudimentary wings and spend their entire life underwater except to mate briefly at the surface. To breathe via their spiracles, they use an air reservoir (in its more sophisticated form known as a 'plastron gill', as in the Diving Bell Spider), here taking advantage of hydro-repellent hairs that trap a thin film of air. This type of gill allows oxygen to diffuse into the trapped bubble from the surrounding water. Only when the bubble becomes oxygen depleted or shrinks due to the loss of nitrogen must it be recharged by trips to the water's surface. The winged Water Veneers always mate on or beside the water surface. The case-bearing larvae of a few Hawaiian aquatic cosmet moths of the genus *Hyposmocoma* have the extraordinary ability to breathe through their cuticle without the use of gills, but being amphibious they can also breathe in the same way on the surface, whereas when in water the currents must be fast enough flowing to deliver sufficient oxygen. The silken cases of *Hyposmocoma* (shaped like 'burritos', 'bugles' and 'cones'; p. 134) make them the caddisflies of the moth world. Their larvae moult underwater. Daniel Rubinoff discovered that *Hyposmocoma* has colonized aquatic environments three times in the last six million years, each species being unique (endemic) to a particular volcano in Hawaii. A final striking example is the erebid tiger moth larvae of *Paracles laboulbeni* (and other species like *P. klagesi*) that feed on submerged plants in slow moving water and ox-bow lakes of South America. Rather than having gills, they have a water repellent cuticle and their dense hairs form a plastron gill, although little is yet known about them.

CAVE MOTHS

Caves are another unusual environment for moths to inhabit. Few moths live permanently in caves. There are tineids (Tineidae) feeding on bat or bird guano, and although some of them are really only known from caves, others are opportunists. Gaden S. Robinson even noted that some of these species had somewhat longer

ABOVE A loose aggregation of Tissue Moth (*Triphosa dubitata*), one of a number of species that can be found overwintering (hibernating) or 'oversummering' (aestivating) in a cellar, outhouse or cave.

ABOVE RIGHT The Herald Moth (*Scoliopteryx libatrix*) is one of the more familiar moths that overwinters in caves, here among limestone in England's Peak District.

appendages and appreciably long antennae as a typical adaptation to cave life, while *Tinea microphthalma* has more reduced eyes than its outdoor living relatives. In the Neotropics, according to Angel Viloria, there are also tineids that feed on seeds that have been regurgitated by oil birds and have been lying for hundreds of years on the cave floor. AZ discovered populations of the Muslin Footman (*Nudaria mundana*; Erebidae) that spend their whole life cycle inside caves, their larvae feeding on a thin film of algae that develops on the moist surface of the rocky walls. Some Hawaiian species of *Schrankia* (Erebidae) studied by Matthew J. Medeiros and collaborators regularly associate with lava-tubes and also show signs of adaptations to underground living, exhibiting flightlessness and wing reduction. Their larvae even feed on tree roots breaching the ceilings of the lava-tubes, just as do populations of the erebid moth *Orectis proboscidata* that were observed by Salvatore Bella on Mt Etna in Sicily, Europe's biggest volcano.

More frequently observed are the species using caves as places where they hibernate or aestivate as adults, such as the tissue moths, *Triphosa dubitata* and *T. tauteli*, the Herald (*Scoliopteryx libatrix*), the Bloxworth Snout (*Hypena obsitalis*), Bogong Moth (*Agrotis infusa*), and also the alucitid *Alucita huebneri*. Cave environments, in fact, can provide more even temperatures than found outside, both in summer and winter. The plume moth *Adaina bolivari* has also been found in caves in Venezuela. It is thought that drawings in the tombs of Egyptian high priest Nesi and the Pharaoh Nebwenenef might represent hibernating plume moths. Apparent depictions of alucitids appear not only in Egyptian tombs, but in the Eland Cave bushmen rock paintings in Namibia, as has been documented by Jeremy Hollmann. It is intriguing that ancient peoples perhaps just painted what was around them, but evinced an aesthetic fascination in Africa for the beautifully divided wings of plume moths.

COLD AND HOT

A few Lepidoptera live in really extreme conditions. *Gynaephora groenlandica* is such an 'extremophile' moth living in Greenland and the High Arctic that can tolerate temperatures below -70°C (-94°F). This lymantriine erebid moth, whose life cycle can take up to 15 years, is a true cryogenic insect, which uses glycerol and trehalose-based anti-freezes in its haemolymph and the caterpillar basks in sunshine in the few available weeks of summer so that it can feed, albeit very slowly (see p. 48). Above 5,000 m (16,400 ft), few families of moths are recorded: Brachodidae, Pyralidae, Geometridae, Noctuidae and Erebidae. A species in the last family, *Palearctia hauensteini*, has been seen at 5,200 m (17,060 ft) on Everest. High elevation records come particularly from the Himalayas, and the high Altiplano and Paramo of the Andes. Many of these records are for day-flying species, but surprisingly, Laszlo Ronkay has documented furry moths flying above 4,000 m (13,100 ft) in the snow at night near the top of deeply dissected Himalayan valleys. Such furry species include the nocturnal genus *Dasypolia*, recorded up to 4,877 m (16,000 ft) in Nepal. Axel Kallies reported the little bear moth, *Brachodes flagellatus* (Brachodidae) in Tibet at 5,100–5,300 m (16,730–17,300 ft). This is perhaps the highest reported elevation for a moth, although there are unverified reports of Everest explorers having seen both butterflies (e.g. Painted Lady) and unidentified moths up to 6,400 m (21,400 ft). It seems unlikely they can breed there. Among minute moths, an adult pigmy moth, *Stigmella nivea*, was collected by Ole Karsholt, flying between snow flurries at 4,700 m (15,420 ft) in the Peruvian paramo. It should be noted also that moths can be carried live at high elevations in the airstream. The Gypsy Moth (*Lymantria dispar*) has been documented as being carried as larvae by winds over a distance of 180 km at elevations around up to 4,267 m (14,000 ft). A wider range of moths can live happily at higher elevations on high mountains in the tropics, and go up to 4,875 m (16,000 ft) in the Ruwenzori. Some Hepialidae can survive elevations as high as 5,000 m (16,400 ft) where the caterpillars and pupae are protected up to 15 cm (6 in) deep underground. Apart from antifreeze in caterpillars, adults of some day-flying moths, such as the Broad-bordered White Underwing (*Anarta melanopa*), Black Mountain Moth (*Glacies coracina*) and Black Mountain Pearl (*Metaxmeste phrygialis*) occurring in high mountains or at high latitudes, use melanin (black pigment) in their wings as well as furry bodies to help thermoregulate and protect against UV rays.

At the other extreme, some moths are adapted to extremely hot environments. A number of species have been recorded from Death Valley, amongst them *Euproserpinus phaeton* (Sphingidae) (p. 148), which emerges very early in the spring and completes its life cycle before the hottest part of the year.

ABOVE As in some other high montane species, the Black Mountain Moth (*Glacies coracina*) is not only day-flying but has a black furry body to absorb the sun's heat at high elevations.

AROUND THE WORLD

Moths not only occur in a wide range of environmental conditions, they also have colonized every continent on the planet except Antarctica. The most northerly resident moth populations are those of the aforementioned *Gynaephora groenlandica* and its kin *G. rossii*, recorded from Northern Greenland and Ellesmere Island at around 83°N. In the Old World, the furthest north reaching resident

BELOW The larva of the Arctic Woolly Bear Moth (*Gynaephora groenlandica*), here on Baffin Island, can usually only feed a couple of weeks a year, its development taking up to 15 years. This caterpillar can occur close to the 80th parallel, and antifreeze in its haemolymph means that during hibernation it is tolerant of temperatures as low as -70 C.

ABOVE A female of *Plutella polaris* was found in 2015 on a south-facing slope at 79.3˚N Svalbard, Norway, 142 years after its original discovery.

moths are *Matilella fusca* (Pyralidae), *Apamea exulis* (Noctuidae) and *Plutella polaris* (Plutellidae) from Svalbard (Spitzbergen). The last species, long known only from a single collection of seven specimens made by the explorer Rev. Alfred Edwin Eaton in July 1873, has eventually been rediscovered in 2015 in one of Svalbard's most northerly fjords, at 79˚20' N. From Severnaya Zemlya, the looper *Psychophora cinderella* has been recorded up to 78°56' N, and the noctuid *Xestia aequaeva* at 78˚37' N. Notably, no moths can occur at equivalent latitudes of the southern hemisphere, where the tip of Patagonia reaches only 54°S. The Diamond-back Moth (*Plutella xylostella*), the world's most widespread species of Lepidoptera, gets blown to such chilly northern latitudes as Severnaya Zemlya and Svalbard. Along with two other migratory moths species, it can also reach the 54th parallel south in South Georgia Island. Campbell Island (52nd parallel south) has about 18 moth species, a third of them with reduced wings (brachypterous), including members of the genera *Tinearupa* (Xyloryctidae), *Campbellana* (Carposinidae) and *Reductoderces* (Psychidae). A few moths occur on other sub-Antarctic islands, where they are often important players in the relatively simple ecosystems that occur there. The Marion Flightless Moth (*Pringleophaga marioni*; Tineidae) is considered a vital link in the recycling of nutrients on the Prince Edward Islands; on Marion Island (46°45' S), it thrives in the relative warmth of albatross nests. Also on Marion Island the brachypterous *Embryonopsis halticella* (Yponomeutidae) is the major herbivore, removing a quarter of all grass biomass.

MOTH BIOGEOGRAPHY

Biogeography is the study of the patterns of distribution of organisms on the planet. Every species occupies a particular range that is the geographic area where it is present. The concept of 'range' can equally be applied to broader groups such as genera, tribes, families, and so on upwards along the classification system. Ranges can be continuous or split (fragmented), and broad (for widespread species) or restricted (for narrow endemics). Detailed study of ranges can reveal shared patterns in space and time which may thus reflect the same underlying causes. More importantly, study of particular moth ranges, when linked to reconstructions of genealogical relationships (phylogeny) or to knowledge of age of relevant fossils, can help to understand how moths globally diversified, even back into deep time. Biogeographers combine their knowledge about current ranges with ancient geological events, as well as information about past ecological conditions, including sea levels and climate. They distinguish two fundamentally different processes leading to present-day distribution patterns: vicariance and dispersal. The former is based on the splits of once continuous ranges by the emergence of gaps, which could have physical causes (e.g. oceans or rafting tectonic plates) or ecological in the broadest sense. The latter is based on movement into new areas by the organisms themselves.

As an example of vicariance, the present-day distribution of various genera of jaw moths (Micropterigidae) is thought to reflect their likely ancient occurrences on tectonic plates. These plates are the pieces of the Earth's crust that have potentially carried their fauna and flora partially intact during continental drift. Those plates that have not actually disappeared under one another have undergone a dynamic dance through time, leading their relative positions to change greatly. This drift is an ongoing process that can be measured across some active fault lines like the African Rift Valley, at a speed of centimetres per year. Of course, most of the original passengers on these tectonic fragments have long since become extinct but several, somehow, have survived. For example, around 180 million years ago, the great southern continent known as Gondwana started to split up into separate land masses, notably South America, Africa, Madagascar, India, Australia, 'Zealandia' (New Caledonia + New Zealand) and Antarctica. Micropterigidae, which preserve well as fossils in amber, are a remarkable living testament to the mostly lost world of Cretaceous moths. These beautiful often gold-and-purple metallic moths have limited powers of adult dispersal and many are denizens of the understory of primary forests, where they swing-hover over damp logs or under ferns in dappled sunlight. They have a rich fossil record, with the most exquisite fossils, preserved in amber, dating back to 130 million years ago in the mid-Cretaceous. Different micropterid genera that belong to a single ancient group, one of five lineages within the family, now occur independently in almost all of these areas, India and Antarctica excepted, but we know from fossils in Burmese amber that they also occurred in the Indian region, Antarctica is nowadays inhospitable to all moths because of deteriorating climatic conditions a few tens of millions of years ago, but it would have had a rich moth fauna.

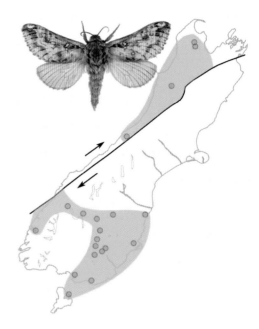

ABOVE The ghost moth *Wiseana jocosa* illustrates the type of distribution expected for the geographic or geological separating process called vicariance. It is conceivable that such a primitive moth with poor dispersal powers might have a distribution reflecting the gradual shearing of its original populations along the Alpine Fault of southern New Zealand.

Several other groups of moths, too, are apparently relicts of a Gondwanan-type distribution, often though absent in a few of its fragments. These include Castniidae, which today occur only in Central & Southern America, Southeast Asia and Australia. The Palaephatidae show an even greater disjunction, nowadays known to occur only in southern South America and Eastern Australia. Another example is the family Sematuridae, which might reflect the break up and separation between South America (their main distribution area, although they have spread northwards as far as Arizona) and Africa (with a single species, *Apoprogones hesperistis*). Such fragmented ranges, especially when supported by evidence of antiquity corresponding to geological knowledge (e.g. via fossils, shared lineages of high taxonomic rank, or molecular sequencing) happen to be very informative in biogeography. They show past links between land masses via the moth groups which are shared by them. The best examples are found in groups for which there is no evidence that they tend to disperse successfully between the areas where they are found today.

Vicariance occurs not only via the major moving plates on the planet but at a fine scale. Geological faults enable populations of relatively sedentary moths on opposing sides of the faults literally to slide past each other. In New Zealand, the

Alpine fault is a strike-slip line that runs down the spine of South Island. George Gibbs has related how two most closely related microphterigid species, *Sabatinca aemula* and *S. chrysargyra*, seem to have evolved vicariantly by the sliding of an ancestral population, at 2–3 cm per year, along this active disjunction, and today the two species nowhere overlap. Populations of the swift moth *Wiseana jocosa* are likewise now separated into northwest and southeast components that might have resulted from such a vicariance. Of course, populations can themselves spread at a much faster rate than landmasses. But DNA methods can be used to show that the locations of populations of certain primitive moths are likely to have been stable in location over long time periods relevant to such geological shifts.

Fundamentally important in biogeography are species or groups exclusive to a given geographic area. These are called endemics. At a high level of classification, occurrence of endemics is particularly striking. In fact, quite a number of families are endemic to a specific land mass or region, e.g. Andesianidae and Mimallonidae (North, Central and South America), Metarbelidae (Africa), Cimeliidae (Mediterranean), Ratardidae (Southeast Asia), Anthelidae and Carthaeidae (Australo-Papuan Region), Whalleyanidae (Madagascar) and Mnesarchaeidae (New Zealand), to mention just a few. Excitingly also, new families are still being discovered, such as the Aenigmatineidae, known only from just off the coast of southeast Australia with the Kangaroo Island

RIGHT One of the amazing examples of endemism, in this case in the Mediterranean region is provided by the Gold moth family Cimeliidae, which is related to hook tip moths. Here, is the stunning gold moth *Axia margarita*.

Moth (*Aenigmatinea glatzella*). For a moth biogeographer, it is a great time to be alive.

A high percentage of endemics usually indicates isolation of that area in relation to other lands. The high number of endemic moth groups known from the Australian region or Madagascar for example (where more than 80% of species might be unique) correlates with the long isolation of these landmasses. This followed the break-up of the ancient southern continent of Gondwanaland. Deep intervening seas between continental fragments meant that land passages were no longer possible. Madagascar is quite close to Africa (minimum 400 km) but it retained great faunal integrity, probably because no land bridges existed for the last 85–125 million years. The dispersal of moths from Africa has been hindered by prevailing trade winds and cyclone tracks from the east. By contrast, long isolated faunas now with strong land connections (notably that of South America, which fused with lands to the north over the last few million years), show substantially lower values of endemism in relation to Central America. Glaciations also played a substantial role either in uniting or exposing land masses. Especially in the case of shallow intervening seas (the case for example for Borneo with mainland Southeast Asia), melting of ice sheets during warm interglacial periods isolated lands. By contrast, lowering sea levels as water was incorporated back into ice sheets, reestablished connections between land masses. Such intervals were often too short to develop really unique moth faunas.

The presence of particular species in an area does not necessarily indicate that they have always lived there and passively rafted on moving landmasses. In fact, moths do disperse, and aerial dispersal has likely been the most important factor shaping the current distribution. Dispersal can occur over vast distances and occasionally a pregnant founder female arrives and is lucky enough to find a viable site to lay her eggs. Many volcanic islands, such as some of the Galapagos or the Hawaiian Islands, did not exist earlier than 5 million years ago, nonetheless they have been colonized over this geologically rather short time by several moth species. A number of these underwent explosive differentiation in such isolation and speciated in what are now numerous insular endemics (see *Hyposmocoma*, p. 135). Every brave new species or founder of an entire lineage has to strike out from somewhere. The existence of some ranges that extend over vast land areas or straddle more than one continent are the result of dispersal by moths from an original source population. A few moths are now truly cosmopolitan in that they occur on all continents except Antarctica.

Dispersal tends to overwhelm or obfuscate ancient patterns of distribution. However, entire families of primitive moths (e.g. Neopseustidae, Acanthopteroctetidae) still show an extremely relictual distribution around the Earth. Such moth distribution patterns are likely to be a result of ancient wider ranges which have now contracted

ABOVE *Mydrodoxa sogai* is a spectacular pantheine noctuid from Madagascar. The species is only known from this single individual, providing an example of an exceptionally rare Madagascan endemic species in a genus also unique to the island.

RIGHT A simple scheme for
the six principal biogeographic
(zoogeographic) regions of the
World. Two pairs of regions are often
combined, as indicated by arrows.
The latitudinal boundaries between
the regions are blurred, since there
is no precise demarcation, but the
moth faunas of each region are
fairly distinct and can often be
further subdivided (e.g. Madagascar
and New Zealand are often treated
separately). Significant parts of the
southern hemisphere landmasses
formed a single one, Gondwana,
in the Jurassic, whilst most of the
northern hemisphere was joined up
into Laurentia.

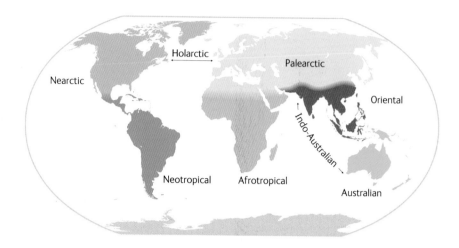

RIGHT A simple scheme for the six principal biogeographic (zoogeographic) regions of the World. Two pairs of regions are often combined, as indicated by arrows. The latitudinal boundaries between the regions are blurred, since there is no precise demarcation, but the moth faunas of each region are fairly distinct and can often be further subdivided (e.g. Madagascar and New Zealand are often treated separately). Significant parts of the southern hemisphere landmasses formed a single one, Gondwana, in the Jurassic, whilst most of the northern hemisphere was joined up into Laurentia.

into smaller and climatically more stable pockets (sometimes known as refugia). For example, new acanthopteroctetid species have recently been discovered in small and ancient enclaves around the world, such as the Brandberg Massif in Namibia.

From the comparative study of biogeographic patterns of moths between different regions, a discipline known as faunistics, measures of similarities and differences between faunas can be assessed. A simple scheme, reflective of the largest scale faunal affinities across the world, shows the regions we have referred to throughout the book. These principal biogeographic regions hold up for many different groups of terrestrial organisms and are mostly separated by large sea gaps. There are transition zones between some of these regions too, for which subregions can also be defined. All are based on patterns that recur between independent evolutionary histories of various moths. Finally, we highlight two major aspects of dispersal: moths that naturally migrate and moths than have recently dispersed around the planet thanks to man.

MOTH MIGRATIONS

Individuals or swarms of moths can migrate incredible distances, and these feats are not always related to the size or power of a moth species. Moths can occur in the mid-ocean and there are many records of moths coming to boats far out to sea. Some 14 species, including *Achaea janata* and the Diamond-back Moth (*Plutella xylostella*) have colonized Easter Island, 2,075 km (1,290 miles) from Pitcairn Island, and 3,512 km (2,180 miles) from the nearest mainland in Chile. *Plutella xylostella* is often considered to be the World's most widely distributed dispersive moth. Feeding on the cabbage family (Brassicaceae), on which it is a pest, singletons have been tracked covering 3,000 km (1,870 miles) and can easily be carried on an airstream considerably more than 800 km (500 miles) in a day. A continuous migration of this species lasting a few days exceeding 3,680 km (2,290 miles) was reported by Roy A. French in 1967. Migrating such great distances is favoured if a moth can be active by day and by night, although air currents undoubtedly help propel such small individuals. However, Massimo Di Rao once observed the crambid *Nomophila noctuella* alighting repeatedly and taking off from surface of the Mediterranean

Sea during calm weather. Nocturnal moths can migrate over land for several nights. Bogong moths migrate over 1,000 km (620 miles) to aestivate in the Australian Alps. *Urania* moths, when in migratory phase, generate huge irruptions of adults which can pass by for days on end. Sedentary individuals are day-flying, but migrating ones also travel at night, *U. fulgens* having being caught as far away from the Central American mainland as Kingston in Jamaica. The Sunset Moth (*Chrysiridia rhipheus*) similarly can come to light when on migration and can be sighted sailing over the summits of mountains far from its hostplants. The eastern rainforest of Madagascar produces outbreaks of sunset moths. At the end of the dry season they may cross the barren plateau of Madagascar in search of restricted stands of their host plant *Omphalea*, already in flush growth in the limestone landscape of the west and northwest of the island. Similar trajectories occur every year in temperate zones, particularly when warm air currents come up from the south. From spring to autumn, a plethora of moths leave Mediterranean countries and spread northwards all over Europe. Strictly speaking, these long-range movements are not true migrations as they do not follow regular paths and no return flight has been observed, not even by the following generation. Such wandering moths are known as vagrants and produce offspring in the new territories they reach. However, the northernmost ones perish in the winter. Despite such a bleak outcome, vagrants arrive every year. Their strategy likely evolved as one to send propagules everywhere. This gives vagrant species a better chance to survive in the long term, should ecological conditions in their primary range become less favourable, or climate becomes more suitable in places they reach.

ABOVE *Urania* moths (here, *Urania leilus* males) mudpuddling. On migration, large aggregations of these moths appear.

INVASIVE SPECIES

Invasive species, sometimes known as 'aliens', are ones that establish in areas where they did not naturally occur. In today's world, moth invasions are a growing menace to natural ecosystems, fields, parks and gardens. They are often inadvertently promoted by trade, e.g. in ornamental plants, and range expansions are frequently facilitated by climate change. Many invasive moth species turned out as pests of stored grain, horticultural trees and shrubs, and crops. In the case of the Diamond-back Moth, its long-distance dispersal has been greatly assisted over several centuries by the expansion of agriculture – specifically the planting of crucifers.

The scale of invasion by moths is probably underestimated, especially in tropical faunas. It is striking that more than 1% of the European Lepidoptera fauna is now classed as invasive and most of these invasives have come from outside Europe. A case of an invasion in Europe, studied by DCL, is the Horse-Chestnut Leaf-miner (*Cameraria ohridella*), which was first noticed in Macedonia in 1984, and described as a new species two years later. As a genus new to Europe it was assumed to come from elsewhere. A high diversity of mitochondrial DNA variants in the southern Balkans, along with the discovery of accidental pressings in herbarium specimens from central Greece dating back to 1879, have shown that this moth is really a unique European native species that was long trapped in a biogeographic enclave. It probably spread around via the transportation of fallen leaves, but principally by car. Such artificial transport enabled the moth to spread like wildfire, colonising most of Europe in just two decades, since its arrival in Vienna in 1989.

RIGHT Larvae of another highly invasive species, the Horse-Chestnut Leaf-miner (*Cameraria ohridella*) are often so abundant that several share the same leaf mine of European Horse-chestnut (*Aesculus hippocastanum*). This species 'escaped' from its enclave in the Southern Balkans in the mid-1980s.

LEFT A recent invader of Europe is the Box Moth (*Cydalima perspectalis*), a crambid inadvertently introduced from China. Both wild stands and urban hedges of European Box (*Buxus sempervirens*) are being severely damaged by this species. In some places densities of this moth are so high that people have to turn their lights out at night.

Indeed, invasive species are arriving in new countries all the time. Just in 2016, the tiny heliozelid moth *Antispila treitschkiella* was first detected in a moth trap in the Natural History Museum, London. The species has turned out to be a prolific leaf miner, spotted on every Cornelian Cherry (*Cornus mas*) examined by DCL from London to Cambridge. It apparently went unnoticed for years. Another invader, the beautiful Box Moth (*Cydalima perspectalis*; Choreutidae), originating in temperate China is becoming one of the more abundant moths in European cities, while the Fig-tree Skeletoniser Moth (*Choreutis nemorana*; Chorcutidae), coming from the Mediterranean, is now stripping leaves of Common Fig (*Ficus carica*) in London. Invasive species confound classical faunistic patterns. For example, the arrival of the beautiful Palm Borer Moth (*Paysandisia archon*) in southern Europe means that the biogeographic range of the family Castniidae is now vastly extended. These moths bore into young fronds of many palm species and are already having a massive impact on the native Dwarf Fan Palm (*Chamaerops humilis*) along the Mediterranean coast.

At the same time that invasive moths are on the rise, populations of native moths are greatly in decline for other reasons, in response to the dramatic changes in the environment caused by humans. Moths are also on the move in response to trends of increasing average temperatures. Native species are on the decline (p. 195) due to loss of their habitats and replacement by monoculture, disruption of their ecology (e.g. changes and movements in parasitoid populations), effects of fertilizers decreasing the diversity of their foodplants in favour of more thuggish plants, widespread use of pesticides, changes in atmospheric gases and water levels, and many other factors. Moths are a fantastically sensitive group to use for monitoring the changes happening in our environment, providing baseline date for reversing as many of those changes as possible. In the next chapter, we consider the origin and extinction of species in moths, how they have evolved and continue to do so.

CHAPTER 7

Evolution in action

THE EVOLUTIONARY HISTORY OF MOTHS is still being unravelled, most recently and decisively by the advent of molecular techniques enabling the sequencing of genes. The DNA sequences of a wide diversity of moths, with genetic information carefully aligned across species, collectively enshrine the traces of events (e.g. mutations) in moth evolution stretching back into deep time. Powerful computer programs combine these signals to elucidate statistically the relationships between groups of moths. The more groups we sample and the more data we analyze, the more confidence we have about the evolutionary relationships of moths. Yet even with sequencing of unprecedented numbers of genes, certain parts of moth history remain hazy. Moths have gone through periods of rapid evolutionary and ecological radiation, and determining the exact series of changes through time is a challenging task. Here we focus on evolution as an ongoing process as well as a pattern from the past.

Moths provide excellent examples of evolutionary processes in speciation and diversification. If a single species changes through time we call this anagenesis. This process may either reflect adaptation through natural selection due to shifting environmental conditions, or the random accumulation of changes with no particular adaptive value (so-called neutral evolution). Sometimes anagenetic changes result in very different looking organisms over time, but not necessarily different species. Examples of such transformation series in Lepidoptera are difficult to find though, because of the limited timeframe available to the observers and the poor fossil record for these insects.

We can spot anagenesis more directly if we focus on populations occurring across the range of one species in the field. This is particularly evident in species with two or more genetically fixed colour or shape forms ('polymorphism'), when one form begins to predominate over the other(s). A classic example is that of industrial melanism, as seen

OPPOSITE The Jersey Tiger (*Euplagia quadripunctaria*, left column) is a model moth for evolutionary studies. In the UK and parts of Western Europe its yellow hindwinged form (top left) has been increasing in frequency over recent decades. Acquisition and spread of such a novelty exemplifies anagenesis, i.e. change within a lineage or species. *Euplagia splendidior* (right column, all from Armenia) is its sister species. This species has greenish shiny forewings and only the reddish hindwinged form. The evolutionary split leading to two evolutionary lineages (potential species), *E. quadripunctaria* and *E. splendidior*, is an example of cladogenesis.

WHAT IS A SPECIES?

Out of the many definitions of species which have been proposed, still the most popular one is Mayr's: "Species are groups of actually or potentially interbreeding natural populations, which are reproductively isolated from other such groups."
A population is in turn a group of actually interbreeding conspecific individuals which live in the same area.

ABOVE The Peppered Moth (*Biston betularia*) is surely the most famous case of evolution in action. The melanic form (top) spread greatly since the 1840s in Industrial Britain but has declined there again since the 1950s due to the Clean Air Act. Examples shown are from the George Taylor Porritt collection in the Tolson Memorial Museum.

in the Peppered Moth (*Biston betularia*). During the 19th century, in some European industrial districts, there was a remarkable increase in frequency of the once rarest black form (*f. carbonaria*) over the typical, largely white 'black peppered' form, that in some polluted localities completely took over the latter. It has been supposed that as these moths were mostly resting on sooty bark, the white form could be picked off easily by predators whereas the so-called melanics were far better camouflaged. Recently, there has been debate regarding the factors promoting the shift between the two forms, e.g. whether due to natural selection by predators favouring black vs white individuals or just by random changes of the relative abundance of the forms. Also the moths' active choice of their resting sites prior to predation likely played a role, but the story of the Peppered moth remains one of the most visible examples of anagenetic change. Moreover, the process of evolution is dynamic, and with atmospheric cleanup of pollutants, especially those deriving from coal, frequencies of forms have changed back from mainly black.

Anagenesis is continuously taking place in moth populations and species. Even when there is no evident polymorphism in external features, changes can often be detected after statistically elaborate analyses of shapes and patterns or shifts in gene frequencies via DNA studies. In one of these studies carried out by J.R. Freeland's research team it was shown, after comparing old museum specimens with extant populations, that slight changes in average wing shape by the Garden Tiger (*Arctia caja*), in Britain have been accompanied by a loss of genetic diversity. Major global changes, like the increase in carbon dioxide levels and artificial lighting, are potential drivers of evolutionary changes in moth populations. Behavioural changes are happening too. In 2017, Florian Altermatt and Dieter Ebert determined experimentally that some colonies of the Spindle Ermine (*Yponomeuta cagnagella*) have become more sluggish due to exposure to urban light pollution.

RIGHT The Garden Tiger (*Arctia caja*) is an example of a moth that was common several decades ago in Britain but has declined drastically, accompanied by subtle changes in wing shape and a reduction in genetic diversity.

SPECIATION

When one species splits into two we have speciation. This process occurs via mechanisms that take a long time but sometimes are relatively sudden. The amazing variety of species and forms of Lepidoptera have long been exploited by researchers to study speciation phenomena, and a great deal of modern evolutionary biology is based on this group of insects. A review of all speciation modes and patterns goes beyond the scope of this book; nonetheless, most species are thought to have originated when populations split into two or more separate geographic areas, e.g. when physical barriers such as mountain ranges or water gaps arose. This is called allopatric speciation. Once geographically isolated, the separate populations independently undergo anagenesis and develop their own evolutionary novelties, which can no longer be shared with their counterparts because of the physical barriers in place. Whilst members of each population become more and more genetically harmonized and attuned in reproductive behaviour, over time the sister populations drift genetically apart to the extent that their members are interfertile no more, if and when they happen to enter into contact again with their former kin. In such areas of secondary contact, hybrids may occur for several generations. However, the harmonized gene combinations that have evolved in isolation are disrupted in these individuals. This may in turn disrupt developmental processes or negatively affect the hybrids' overall physiological and ecological performance. The hybrids are then not as fit as the offspring from each population and therefore face a selective disadvantage. Any characteristic that prevents matings between individuals of the two populations would hence be favoured by natural selection, and ultimately refine the speciation process. We know, however, many cases of stable, geographically narrow hybrid zones where this final step has not been achieved.

An analysis of the distribution patterns of many pairs of closely related species does indeed confirm the predominant allopatric nature of speciation, with or without formation of a narrow hybrid zone. A good example of the latter is offered by the pair of pine hawkmoths, *Sphinx maurorum* and *S. pinastri*. These almost indistinguishable species partition Europe between the southwest (*maurorum*) and north-central-east (*pinastri*) due to spatial isolation during the ice ages. Where they

BELOW The almost indistinguishable pine hawkmoths, *Sphinx maurorum* (left) and *Sphinx pinastri* (right; both females) provide a good example of a species pair occurring in separate areas (having diverged via allopatric speciation), but sharing a narrow overlap zone in southeastern France in which they hybridize.

RIGHT Two burnet moth species provide an example of parapatric distributions: the more widespread *Zygaena rhadamanthus* (left) and its Italy-restricted sister species *Z. oxytropis* (right) have separate distributions that just touch in the Ligurian Riviera.

have regained secondary contact in France, they form a narrow but stable hybrid zone with intermediate individuals. In other species pairs, full reproductive isolation has been achieved so that no intermediates occur in overlap zones.

Spatial isolation between two groups of populations can also develop via elevation, one group specializing in warmer lowland conditions and another in colder highland regions. One such instance seems to be the narrowly diverged species pair of the Small Eggar (*Eriogaster lanestris*) along with *E. arbusculae*. Populations of these eggars in the Alps are displaced elevationally, often with a belt in-between where neither of them occurs. However, the past distribution of their populations with respect to palaeoecological events is still unknown and what we observe today could be the outcome of a different path leading to speciation. For example, it could be a secondary elevational overlap between populations that had previously differentiated by allopatric (geographic) means.

A minor variation of the model of allopatric speciation is peripatric speciation, when a small population buds off from a major parent range of populations.

RIGHT The Small Eggar (*Eriogaster lanestris*) is one of a pair of species distributed along an elevational gradient. *Eriogaster lanestris* is adapted to the warmer conditions typical of low elevations than the more montane *E. arbusculae*.

According to Monroe W. Strickberger, *Samia fulva* (Saturniidae) from Andaman I. is supposed to have originated by budding off from *S. canningi*-like ancestors which colonized this archipelago.

Another model is that of parapatric speciation, which occurs between contiguous groups of populations without the interposition of a physical barrier. This has been shown to occur in a number of other organisms, including grasshoppers, when major chromosomal rearrangements take place between adjacent populations – chromosomes being the cellular structures carrying the genetic material – but it is yet unconfirmed for Lepidoptera. It is also supposed to happen when populations showing gradual variation in some of their features along an ecological and/or geographic gradient (a pattern called a cline) diverge to such an extent that those at the two extremes speciate.

One likely case of ongoing parapatric speciation has been studied by AZ in the South-Western Alps, where typical populations of *Heterogynis penella* from the west gradually transform into utterly divergent ones in the east (in the Ligurian Alps and Riviera). Males of this day-active moth can fly normally, whilst females are wingless and legless. The latter remain permanently bound to their cocoons, from which they partially protrude for attracting males with their pheromone lure (see Chp. 4). Eastwards of Nice, France, the male genitalia become longer and more stylet-like, most so in the surroundings of Laigueglia, Liguria, Italy. There is a reason for that! Caterpillars of eastern females spin outside their cocoon an additional silken layer under which the adult females remain after they emerge. Attracted males have to pierce that external layer with their stylet genitalia to reach the females sitting underneath on their inner cocoon. If normal western males are offered Ligurian females they are fully attracted but cannot pierce the additional layer with their unmodified genitalia. Conversely, western French females, lacking that external layer, can be fertilized by their local males as well as distant Ligurian suitors, were the latter given the opportunity in captivity. As the populations are distributed continuously throughout the whole zone, there is a cline in features directly involved in reproductive isolation (and clines in gene frequencies too). Populations at the extremes of the cline tend to behave as different species – particularly for the western males which face a strong physical barrier to reproduction. Populations at the centre of the cline in the French Maritime Alps have also been subsequently named as an independent species (*Heterogynis valdeblorensis*) by Patrice Leraut. However, these represent intermediate ones whose male genitalia show moderate elongation and female cocoons exhibit an imperfectly woven, fluffy and fairly loose external layer.

Finally, when a species originates out of an ancestral one within the same area of distribution, the speciation is said to be sympatric. A notable example of sympatric speciation is the case of the Orchard Ermine (*Yponomeuta padella*) and Apple Ermine (*Y. malinellus*), two superficially identical ermine moth species that share the same range in Europe. *Yponomeuta padella* feeds on a wide variety of trees in the family Rosaceae: blackthorn, plum, pear, cherry, hawthorn and occasionally also apple. In contrast, *Y. malinellus* feeds primarily on apple trees.

BELOW A bivouac of caterpillars of Apple Ermine (*Yponomeuta malinellus*). This moth is practically confined to apple whereas the Orchard Ermine (*Y. padella*) has a broader diet on other Rosaceae trees and shrubs especially Blackthorn (*Prunus spinosa*) but avoids apple. The two species form distinct hostplant races and are thought to have speciated sympatrically and/or ecologically.

It has been suggested that the latter speciated from *Y. padella* by specializing on and essentially exploiting apple. In the case where the two species occur in the same place, *Y. padella* avoids attacking apple trees, possibly since the host specialist *Y. malinellus* would outperform it. Even when the latter is absent though, *Y. padella* shows a really low preference for apple. A third closely related species, the Spindle Ermine (*Y. cagnagella*) is thought to have evolved sympatrically from *Y. padella* by a reversal to a more traditional ermine moth hostplant, Spindle (*Euonymus europaeus*). Spindle Ermine females still retain their ability to respond to chemical cues in Rosaceae leaves. Incipient sympatric speciation through host race formation has also been studied by Igor Emilianov and others in the bud moth *Zeiraphera griseana* (= *Z. diniana*), whose spruce and larch races are partially isolated reproductively but not distinguishable by DNA barcodes.

WHAT IS SO FASCINATING ABOUT MOTH GENITALIA?

It might seem strange, but despite the kaleidoscopic patterns of moths' wings, lepidopterist researchers are usually more interested in what the reproductive organs look like. Indeed, checking for similarities and differences in the shape of genitalia is an excellent way of identifying a species. As we have seen before, a single species may be tremendously variable superficially, while multiple species can be indistinguishable at first glance. In both cases the genitalia usually hold the key. One example that is perplexing for people examining moth-trap contents is identifying the confusingly similar species belonging to the genus *Cnephasia* (Tortricidae*)*. Moth genitalia provide useful clues for clearing up any ambiguity over an individual's identity. There are two main reasons for this.

First, as some parts of the genitalia have to interlock firmly between males and females to ensure insemination, distinct albeit similar species often have different 'lock-and-key' mechanisms so as to preserve their genetic integrity (or at least, avoid frivolous and wasteful liaisons). Genitalic differences in these tightly complementary parts of males and females tend therefore to be clear-cut even between closest relatives, and their examination may soon reveal the true identity of their bearers, given good reference keys, illustrations or anatomical preparations to compare.

Second, other parts of the genitalia just function for an embrace, i.e. generalized clasping between male and female. This is seen in rather rare but mistaken, unproductive pairings between different species. These not strictly interlocking parts evolve independently in each species but they seem to be less driven by natural selection, compared with wing pattern traits. Closely related species may therefore look extraordinarily similar externally just because of the camouflage needed to rest on a similar substrate, or because their patterning is constrained by mimicry (see p. 123). For each species, changes in the shape of these genitalia parts will accumulate with time, which allows us to distinguish them. Even supposedly well-studied species in the northern hemisphere have recently been split into two or more sister species as revealed by an examination of their genitalia.

Some of the best known examples of species pairs in Europe are the Transparent burnets, *Zygaena purpuralis* and *Z. minos* (Zygaenidae); the Broad bordered yellow underwings, *Noctua fimbriata* and *N. tirrenica*; the Common and the Lesser common rustics, *Mesapamea secalis* and *M. secalella*; and the Grey and the Dark daggers, *Acronicta psi* and *A. tridens* (here with a third species, *A. cuspis*)

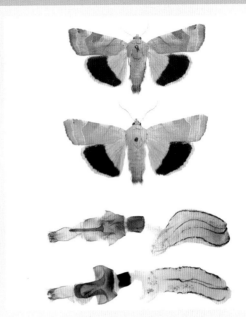

ABOVE The Broad-bordered Yellow Underwing (*Noctua fimbriata*, top moth) and *N. tirrenica* (bottom moth) are two externally hard to distinguish sister species which broadly overlap in southern Europe. Both males and females can, however, very easily be separated by their genitalia. Here the female of *N. fimbriata* (top genitalia) has a finger-like rather than tongue-like process as found in *N. tirrenica* (bottom genitalia).

(Noctuidae). Sometimes, however, differences in the genitalia are not easy to evaluate and sophisticated statistical analyses of numerous samples (or DNA data) are required to assess whether they are really significant.

Strongly similar sister species that can hardly be distinguished on a morphological basis are said to be sibling species. These can usually be distinguished by differences in biological traits such as different hostplants, flight or mating time and pheromone blend. This is easier to assess if the sibling species coexist, then excluding that differences simply reflect geographic variation within a same species. A striking example of sibling species looking exactly alike, even in the genitalia, is that of *Apamea michielii* and *A. maillardi* (Noctuidae), the former substituting the latter over a large sector of the southwestern Balkan Peninsula. We know that they are two species because the first lacks the male pheromone brushes that the second possesses, so their courtship behaviour is certainly different. Sibling species may also have different chromosome numbers and will always show at least some genetic differences.

THE TREE OF MOTHS

A great race took place between teams of molecular researchers to come up with the first big phylogenetic tree of Lepidoptera. In 2010 a Finnish team led by Marko Mutanen and Lauri Kaila trounced the goliath 'LepTree' team in the USA led by Jerome C. Regier to publish the first big analysis covering most of the Lepidoptera lineages. Since then, a flurry of molecular papers has been filling in gaps in the story, piece by piece, some integrating salient morphological characters. Some parts of the complete moth radiation remain extremely difficult to crack. Meanwhile, the major evolutionary branches within the Lepidoptera are only now starting to come into focus (see p. 4).

The exact timing of this evolutionary story is dependent largely on plotting accurately dated fossils on the tree, but finding moth fossils sufficiently well preserved to show the key morphological features is exceptional. Here, we paint a broad picture of our current knowledge of moth relationships, a picture that may change significantly in the near future. Experts currently recognize seven families of butterflies in addition to some 130 extant moth families. An additional four moth families are known only from the fossil record, mostly from impression fossils of exceptional quality found in the Jiulongshan Formation of Inner Mongolia, northeast China. These fossils date from the Middle Jurassic, around 165 million years ago.

THE FIRST MOTHS

Among the earliest known moths is *Archaeolepis mane*, an impression fossil discovered in calcareous rock from Dorset, UK, dated to around 190 million years ago in the Early Jurassic period. Described by Paul Whalley in 1985, this has since been celebrated as the most important fossil moth find in Britain if not globally. A scanning electron microscope picture has revealed imprints of clear Lepidoptera scales on its wings. Jörg Ansorge documented a few supposed Lepidoptera fossils of similar age in Germany. These were also impression fossils in fine grained Lias (limestone) rock. Recently, detailed analysis of pollen cores from Germany has revealed an amazing diversity of loose scales, some of which look like those of extant primitive Lepidoptera. These cores date back to the Early Jurassic/Late Triassic, about 190 to 225 million years ago. Lepidoptera belong to the superorder Amphiesmenoptera (see p. 7), which also includes caddisflies (order Trichoptera) and a fossil group newly discovered by Wolfram Mey from Burmese amber (order Tarachoptera). Both Trichoptera and Lepidoptera spin silk as larvae, and all three orders sometimes have hairs modified into scales, although only Lepidoptera have wings and bodies that are entirely covered in scales (or occasionally with transparent areas).

The earliest split in the evolutionary history of Lepidoptera that we can reconstruct using molecular techniques is between the so-called jaw moths

BELOW The greatly magnified wing scales of the earliest known fossil of a moth in Britain, *Archaeolepis mane*. This impression fossil was found by a lucky fossil hunter, J. F. Jackson, in Charmouth, Dorset, and is around 180 million years old.

(micropterigids) and all other Lepidoptera. Members of the Micropterigidae retain primitive features, including their wing venation and fully functional mandibles. Jaw moths date back to the time before the iconic proboscis of the Lepidoptera evolved. Adult micropterigids are almost all day-flying, sporting glittering metallic reflections. There are probably more than 260 species globally. They have not colonized young volcanic islands and, where they do occur, each genus is essentially tied to a main tectonic plate or a relictual fragment of it. Some of the richest places in the world for them are New Caledonia plus New Zealand (remnants of the Zealandia plate) and Madagascar, as well as Eurasia. They are so far unknown from the large expanses of lowland tropical rainforests in Africa and South America, and they should be searched for in these areas as well as likely places such as southern India and Sri Lanka (they have recently been discovered in western Australia).

Another very primitive moth lineage is the caddisfly-like genus *Agathiphaga* (Agathiphagidae). As the name suggests, its larvae feed on cones of Kauri trees (*Agathis*). The first species was discovered in Queensland, Australia, by Lionel Dumbleton in 1952. Later, another species was discovered in Fiji, Vanuatu and the Solomon Islands. According to Murray Scott Upton, the larva of *Agathiphaga* can survive up to 12 years in a Kauri cone, perhaps allowing for survival and dispersal when they float on oceans, explaining the wide Pacific distribution of the genus. These two families belong to the superfamilies Micropterigoidea and the Agathiphagoidea respectively; in adults of the latter, the large mandibles are used only during eclosion.

The next offshoot in the moth tree consists of the superfamily Heterobathmioidea, in which there are also functional mandibles in the adult. These are glittery moths that mine the leaves of southern beeches, *Nothofagus,* in Chile and Patagonia.

Next we have the split between the above 'tongueless' moths and those with a proboscis (comprising the suborder Glossata). Acquisition of the proboscis was a key step in moth evolution because it allowed them to sip water and other liquids. This

RIGHT *Eriocrania cicatricella* represents another primitive moth family, Eriocraniidae, confined to northern temperate regions. This is the earliest diverging extant lineage following the development of the proboscis in Lepidoptera. These moths do not seek nectar but sometimes drink dew and their larvae mine leaves of trees such as birch.

BELOW A Kangaroo Island Moth (*Aenigmatinea glatzella*) sits on its gymnosperm hostplant Southern Cypress-pine (*Callitris preissii*). This moth with its strange rather naked head belongs to a newly discovered family of primitive moths, Aenigmatineidae, whose larvae live in *Callitris* cones. The first examples were found in 2009.

step is marked by the appearance of sparkling day-flying moths known as Purples in the superfamily Eriocranioidea. The adults have rudimentary mandibles that assist escape from the cocoon, thereafter they are non-functional; they also feature a short proboscis to suck dew or sap. Purples can be found in the early spring-flying around birches and oaks in temperate parts of the northern hemisphere; their larvae are leafminers with droppings (frass) concatenated in short chains.

The next offshoot is the superfamily Neopseustoidea, now containing three families. Molecular studies have reunited the family Neopseustidae with the leaf-mining and also morphologically highly disparate Acanthopteroctetidae, together with the gymnosperm-feeding Aenigmatineidae, discussed below. Neopseustoids are relics of an ancient global distribution. Acanthopteroctetids occur in places as disjunct as western USA, the southern tip of the Crimean Peninsula, Kyrgyzstan, the Pyrenees, Peru and the Brandberg Massif, Namibia.

The story of the discovery of the Kangaroo Island Moth (*Aenigmatinea glatzella*) is particularly interesting. In 2009 a sharp-eyed ranger, Richard Glatz, noticed, in the southeast part of Kangaroo Island, 13 km (8 miles) off southeast Australia, a glittery purple moth with an odd, semi-naked neck area resting on its hostplant, a tree of the cypress family of the genus *Callitris*. This insect represented a previously unknown lineage of moths and only the second known group of primitive moths to feed on gymnosperms, laying its eggs in *Callitris* cones. The species name *glatzella* is a pun – it refers to the bald head (German: Glatze) of the moth not that of the discoverer! DNA sequencing later confirmed that it merited a new family, Aenigmatineidae, within the superfamily Neopseustoidea.

Next up in the tree (see p. 4) is the superfamily Lophocoronoidea, comprising a few species in Australia. Molecular studies support the idea that these are related to the Exoporia, known for their unique reproductive system, equivalent to the superfamily Hepialoidea (swifts, Hepialidae with four other families and the allied, tufty-headed, New Zealand Mnesarchaeidae).

The next major evolutionary event marks the emergence of the Heteroneura, comprising all remaining Lepidoptera, which have substantially greater differences in the venational patterns between the forewing and hindwing than the Homoneura (all the moth groups discussed above), in which both the wing pairs are similar in venation (see p. 23). Heteroneura split into the superfamily Nepticuloidea (incorporating the two families of eyecap moths, Nepticulidae and Opostegidae) and all other superfamilies. Included here are the smallest as well as some of the largest micromoths, namely the Andesianoidea, which were recently discovered in the southern Andes. These moths, up to about 5.5 cm (2.2 in) in wingspan, were originally deemed to be related to goat moths (Cossidae). Into the lower Heteroneura also fall the Adeloidea (including the yucca moths and the well-known long horn family Adelidae). Progressing further along this stem, we find the leaf-mining superfamily Tischerioidea and the leaf-spinning Palaephatoidea. The last superfamily contains two separate lineages, one from South America and one from Australia. Some of these species feed on trees in the archaic angiosperm family Proteaceae and are related to the vast lepidopteran assemblage Ditrysia (which contains around 30 superfamilies, 29 of which moths, and over 98% of Lepidopteran species richness).

LEFT *Opostega salaciella* is a primitive moth with prominent eyecaps in the family Opostegidae, which share this feature with their closest relatives, the pigmy moths (Nepticulidae).

THE DITRYSIA

Some 150 or more million years ago, the Ditrysia split into Meessiidae proper on the one hand (with the genera *Eudarcia* and *Bathroxena*) and all other ditrysian Lepidoptera on the other. Moving up this part of the tree (lower Ditrysia), we have the superfamilies Tineoidea (comprising four families, including clothes moths and bagworm moths), Gracillarioidea leaf-miners (four families), and Yponomeutoidea (10 families, including the spinning ermine moths). Originating some 10–15 million years later, the so-called Apoditrysia (slightly more advanced Ditrysia) appeared. The lower Apoditrysia are still poorly known as regards relationships of its members (see p. 4). These include the plume moths with their feather-like wings (superfamilies Alucitoidea and Pterophoroidea), the Carposinoidea with their forward-pointing palps, the strangely named Schreckensteinioidea (bristle-legged moths), and the Epermenioidea, showing at rest a tufted silhouette. The most diverse and well-known lower apoditrysian superfamilies are the famous leaf rollers (Tortricoidea), the internally feeding goat and carpenter moths and clearwing moths (Cossoidea), and the glamorous, often aposematic Zygaenoidea, which include the burnet and forester moths. Further Apoditrysia are the false burnet moths (Urodoidea) with their elegant stalked open network cocoons and a recently described Central Asian family, Ustyurtiidae, the often brightly coloured Immoidea, the leaf-skeletonizer moths (Choreutoidea) and the webworms (Galacticoidea).

ABOVE RIGHT A micromoth from China represents the superfamily Gelechioidea, a particularly diverse superfamily containing about 18,500 described species, showing one of their characteristic features, the labial palps which are strongly recurved over the head. This unidentified species belongs to the family Stathmopodidae, exhibiting spurred hindlegs.

RIGHT A Garden Lance-wing (*Epermenia chaerophyllella*) represents the Epermeniidae, one of many families among the lower Apoditrysia. Members of this family often have prominent scale tufts along the dorsal edge of the forewing.

Around 120 million years ago the group Obtectomera appeared. In its lower section this includes the megadiverse superfamily Gelechioidea with their usually upcurved labial palps. The precise branching order is again still to be elucidated, but includes the enigmatic western Madagascan Whalleyanoidea, the Hyblaeoidea (now including the once mysterious East African family Prodidactidae, a relationship that remains to be clearly demonstrated), the Thyridoidea or tropical leaf moths, and the sometimes butterfly-like and entirely Old World Calliduloidea. Finally, the lower Obtectomera include the butterflies themselves or Papilionoidea (including the peculiar neotropical night-flying Hedylidae butterflies), the massive superfamily Pyraloidea (crambids and pyralids), and then the Neotropical, asymmetrically resting Mimallonoidea.

The last superfamily is the closest relative of the recently redefined Macroheterocera, which comprises Drepanoidea (the hooktips and Mediterranean gold moths, Cimeliidae) and then branches off to the Bombycoidea (including the saturniids and sphingids) plus the Lasiocampoidea (the eggars). The next two superfamilies of Macroheterocera are also the largest. Geometroidea has almost 24,000 described species, consisting of the epicopeiids, geometrids (loopers), uraniids and the newly described Southeast Asian family Pseudobistoniidae. The Geometroidea are most closely related to the superfamily Noctuoidea (around 42,500 described species), which comprises the Oenosandridae, owlet moths or Noctuidae, Nolidae, prominent moths or Notodontidae, Euteliidae and Erebidae. This is the largest noctuoid family, which now includes long-established families such as the tiger moths (now subfamily Arctiinae) and the tussock moths (now subfamily Lymantriinae). Some of the day-flying representatives of Noctuoidea are quite spectacular (the noctuid subfamily Agaristinae, for example), though among the most popular are the nocturnal erebids of the genus *Catocala*.

ABOVE *Ischyja inferna*, here feeding on fallen forest fruits, is a large and striking erebid moth which, unusually, exhibits structural colours on the hindwing. Erebidae, containing nearly over 24,500 described species is currently the largest family in the largest superfamily, Noctuoidea (with 42,500 species), one of five among the Macroheterocera.

LEFT *Iotaphora admirabilis* (Geometridae) from Taiwan is one of the most beautiful looper moths, belonging to the also megadiverse Geometroidea (with 23,750 species), which is the superfamily probably most closely related to Noctuoidea.

BELOW Another striking superfamily of Macroheterocera is the Bombycoidea, containing nearly 4,800 species. Among the most beautiful of these from the Southeast Asian tropics is the owl moth *Brahmaea hearseyi*.

It should be noted that some hitherto familiar groupings such as 'Macrolepidoptera' have totally disappeared in the above portrayal of Lepidoptera diversity except to be retained as loose terms. Macroheterocera accurately represent the most advanced Lepidoptera, those that probably evolved since the late Cretaceous era around 100 million years ago. Even the butterflies, classically regarded as a pinnacle of evolution in Lepidoptera, are no longer regarded as 'Macrolepidoptera', but represent an older, around 110 million-year-old group among the Obtectomera. A final scientific consensus of the precise closest relatives of the butterflies among the moths is still eagerly awaited. New families are still being discovered, such as Aenigmatineidae Tunzidae, Ustyurtiidae and Pseudobistoniidae within the last few years. Again, what an exciting period for moth lovers to be alive.

ADAPTIVE RADIATION AND EXTINCTION

The moth tree is a manifold story of adaptive radiation into almost all terrestrial environments. Larvae of the earliest moths, like micropterigids today, likely thrived on soil fungi or exposed on liverworts, while adults fed on fern spores. The appearance of seed-bearing plants (spermatophytes) opened up new opportunities to the early Lepidoptera lineages. Gymnosperms provided resources in early forests which some primitive groups could exploit, like the cones of kauri pines in the case of Agathiphagidae. But it is only with the rapid diversification of flowering plants (angiosperms) during the mid-Cretaceous that moth lineages could really proliferate. Moths did not take long to evolve sophisticated ecological relationships with flowering plants, such as the development of leaf-mining habits in Heterobathmiidae, Eriocraniidae and Acanthopteroctetidae. Even micropterigids could later colonize angiosperms by grazing on live and dead tissues as larvae or eating pollen as adults. However, it was the development of the proboscis that really made a difference in the ecological relationships of adult moths. Flowering plants allowed a much closer association by offering energy from nectar in return for pollination. It also led to a plethora of lifestyles in larvae, such as external herbivory on leaves and flowers, and internal modes, not only leaf mining but also leaf rolling, galling, wood-, seed- and fruit-boring. Despite this mounting association between Ditrysia (p. 172) and flowering plants, however, non-angiosperms were not abandoned. Some moths went on in a major way into detritus and fungus-feeding, some (e.g. callidulids) specialized on ferns. Some of the more extraordinary plant resources recorded are horsetails (in the case of a gelechiid, *Gnorimoschema herbichii*) and Gingko (hosting a few polyphagous psychids, tortricids and saturniids). Others specialized on conifers (pines, cypresses), particularly some families like tortricids. Finally, other more bizarre non-plant resources became exploited, as mentioned in Chapter 3. Among the tineoids in particular, arose fungivory and the ability to break down keratin (i.e. eating feathers, horn and even tortoise shells) and to digest wax in some pyralids. Fungus feeding also occurred among the Erebidae (Boletobiinae), while lichen feeding appeared mainly in the bagworms (Psychidae) and owlets (Noctuidae: Bryophilinae and Erebidae: Lithosiinae). Unlike beetles and flies, however, surprisingly few Lepidoptera eat dung. The radiation of Lepidoptera is thus amazingly complex and not a simple progression coupled with the evolution of plants. There even arose carnivorous larvae that fed on scale insects, cicadas, as well as ant broods, or cohabited with other animals.

On every continent moths have proliferated, except in Antarctica, which is (still!) Lepidoptera-free, even though a few moths elsewhere are known to tolerate freezing conditions (p. 149). Most islands have also been colonized by one or more species. There are no marine moths, but a few larvae can tolerate feeding in brackish pools (*Hyposmocoma*) and some occur in freshwater, living on weeds (notably the acentropine pyralids). Some adult moths, however, can make journeys of many thousands of kilometres by flying across oceans, sometimes aided by transport in the upper airstream; occasionally transport on rafts of vegetation is supposed to have occurred.

Some moth groups underwent massive evolutionary radiations (see Chp. 6) and a single genus may embrace more than 1,000

ABOVE Among the most spectacular radiations in the Southeast Asian tropics is that of the generally poisonous Zygaenidae subfamily Chalcosiinae. The array of colours and forms that has evolved in this biogeographic region beggars belief.

species. The genus *Eupithecia* has almost 1,500 species feeding on a wide range of plant parts and occupying many ecological niches. Evolutionary radiations can and often do occur over relatively short geological timescales of few million years. Some of the most famous that appear to originate from a single ancestor are found in Hawaii. One example is the genus *Hyposmocoma* (see p. 135). From a single presumed founder female, a single lineage of moths has radiated into a breathtaking variety of species (over 350), each with its own ecological and behavioural lifestyles.

Biogeographic regions have their own special faunas, yet it is surprising how many moth genera are shared between the various regions. Increasing human activities such as the plant trade are nowadays spreading numerous invasive moth species like a plague across the planet. Newly invasive species are recorded every year in zones far away from their original geographical ranges, some even perturbing biological communities of pristine habitats in primary forests.

ABOVE *Akainalepidopteron elachipteron* is an extinct Middle Jurassic moth, exquisitely preserved as an impression fossil in the Jiulongshan Formation of China. It represents an also extinct family of Lepidoptera (Eolepidopterigidae). Inset are details of its wing venation.

MOTHS R.I.P.

The general trend in speciation is for the number of species to increase over time. However, extinction is a natural evolutionary process too, which can counteract the tendency for biodiversity to increase over time. Extinctions do not necessarily occur through natural catastrophes such as giant meteorite strikes, volcanic eruptions or even cyclones but also as a result of gradual ecological processes affecting populations or by loss of sufficient genetic diversity to survive in the longterm. Also, when invasive species propagate over a new territory, local species with similar ecological requirements may be unable to compete and can be driven to extinction. We know that entire higher lineages of moths are extinct, as seen in families such as Eolepidopterigidae and the Mesokristenseniidae, preserved only as Jurassic fossils. Of the 229 described Lepidoptera species (of which 199 are moths) known as fossils, only 21 are thought to belong to modern-day species. Islands or small isolated habitats, rather than large continental areas, are more likely to host extinctions.

RECENT EXTINCTIONS

Compared with butterflies, where several undisputed cases of recent extinctions are known, it is pretty hard to point to conclusive extinctions of moth species at a global level. For example, the Frosted Phoenix (*Titanomis sisyrota*) of New Zealand, a large enigmatic tineoid-like moth whose exact systematic placement is still uncertain, is incredibly rare. Only eight specimens are known, the last one collected at the floodlights of the Waipapa Dam in 1959 (sadly the specimen is lost). The preceding specimen was found in 1921. Since this endemic moth is so naturally rare we cannot be sure it is really extinct! To prove extinction convincingly may require exhaustive

RIGHT A female of Frosted Phoenix (*Titanomis sisyrota*), an enigmatic moth of unknown family, which was last collected in the Waipapa Dam in New Zealand in 1959. This species might be extinct, or it might simply just be very rare.

ecological surveys, and this may be particularly difficult for the hundreds of species in museums only known from the single type specimen from the original locality. Any of these are candidates for the rarest moth in the world! Natural rarity is pervasive, as suggested by the number of so-called singletons in large tropical forest samples, which can exceed 50%!

Perhaps the most iconic case of a moth extinction also concerns one of the most beautiful moths in the world – the spectacular Jamaican endemic Sloane's Urania (*Urania sloanus*). Interestingly, the earliest known specimen of this species, possibly collected around 1670, is preserved between mica sheets in the Petiver collection at the Natural History Museum, London, while its last observations were around 1894–1895 (a pair of specimens are known to be dated 1908, but this may not necessarily be the collection date). The naturalist Philip Gosse, writing in 1851, was very familiar with the moth and described in detail its life history and its larvae feeding on the Jamaican Cobnut (*Omphalea triandra*), a plant that is still common in Jamaica. Why did this beautiful insect die out? We do not know. Habitat destruction does not seem to be the explanation, as the hostplant was not eradicated. One wonders if the population reached a critically low level owing to cyclones such as those that hit the island in August 1880, October 1884 and June 1886. Remarkably for an extinct species, fragments of mitochondrial DNA from a museum specimen were recently recovered by Vazrick Nazari and colleagues, who showed that *U. sloanus* is most closely related to the more widespread *U. fulgens*.

Another classic case is the reported extinction of the Fiji Coconut Moth (*Levuana iridescens*; Zygaenidae), a disastrous experiment in biological control. In 1877, the moth had started to become a serious defoliator of coconut palm fronds. In 1925, after unsuccessful attempts to control the species with pesticides, a biological control agent was introduced. This was *Bessa remota*, a tachinid fly, a known parasitoid of a related moth in Malaysia, *Artona catoxantha*. Unfortunately, the control was too successful, and by the mid-1950s the Fiji Coconut Moth could no longer be detected. By the mid-1960s, the same tachinid fly had locally eliminated on Fiji another zygaenid, *Heteropan dolens*, making the introduction of the fly doubly tragic, although fortunately this moth still occurs on Aneityum (Vanuatu I.). The extinction of the Fiji

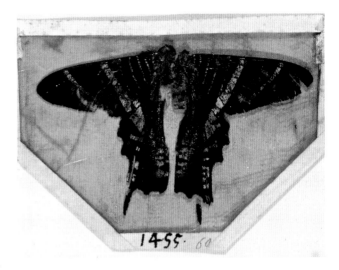

ABOVE One of the oldest specimens preserved in a Lepidoptera collection is a male of the now extinct Sloane's Urania (*Urania sloanus*), sandwiched between mica sheets and probably predating 1700. This comes from the James Petiver collection at the Natural History Museum, London. The moth was last recorded in the wild for sure around 1895.

BELOW One of the world's rarest moths? The presumed original specimen (holotype) of the 'butterfly' *Hesperia busiris* is actually an African agaristine noctuid moth (now known as *Heraclia busiris*). This species has never been rediscovered in the wild and could be extinct. A hole in the thorax indicates that this specimen has been repinned and reset.

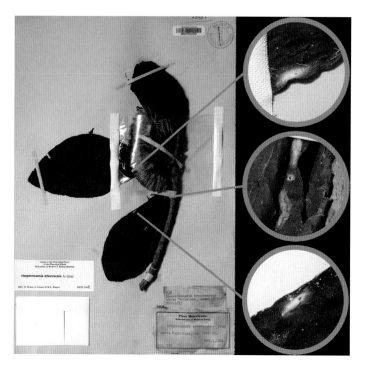

ABOVE The pupal tents of a putatively extinct gracillariid moth, *Philodoria* sp. appear on leaf-mines inadvertently pressed in a herbarium specimen of a critically endangered tree, the Lanai Island-aster (*Hesperomannia arborescens*) from the Hawaiian Islands. Despite detailed searches, the leaf-miner has not been found again.

Coconut Moth has been contested by Mark S. Hoddle, since another population might survive on an as yet unsurveyed island. As in the case of *Urania sloanus*, old specimens of both species have now been successfully DNA barcoded (see p. 140) but nevertheless, moth species can indeed become extinct before they are even properly documented.

The best example of potentially undocumented moth extinctions is from Hawaii. There, some of the beautifully coloured and host-specific gracillariid leaf-miners of the genus *Philodoria* may now only be represented in herbarium specimens, where their early stages were accidentally pressed. The larva of each species makes a distinctive trail or gallery in a leaf of a particular plant species, and it is possible to recognize previously unknown species this way. Several *Philodoria* hostplants (notably *Hesperomannia*) are already down to the very last individuals, if not totally gone. Recently, the mines of an undescribed species of *Philodoria* have been discovered in leaves on herbarium sheets of a tree, *Hesperomannia arborescens*, long since chopped down. In Hawaii various moths are now regarded by the International Union for Conservation of Nature (IUCN) as extinct. These include the Ko'olau Giant Looper Moth (*Scotorythra nesiotes*), the Kona Giant Looper Moth (*S. megalophylla*), and the Hopseed Looper Moth (*S. paratactis*), several erebids such as the Laysan Dropseed Moth (*Hypena laysanensis*), the Hilo Moth (*H. newelli*), the Lovegrass Moth (*H. plagiota*), and at least six noctuid species.

The catastrophic Sweet Chestnut Blight, caused by a fungus introduced from Southeast Asia into North America, which killed 3–4 billion trees of *Castanea dentata* in the first half of the 20th century, is supposed to have totally eradicated several micromoths. These include the Chestnut Ermine Moth (*Argyresthia castaneella*), known only from New Hampshire and Vermont. It is also supposed to have wiped out the Phleophagan Chestnut Moth (*Zimmermannia phleophaga*) and *Ectoedemia castaneae* in Virginia. However, not all is lost, since the last species has recently been shown to be the same as *Z. bosquella*, with a broader distribution. Strange as it sounds, this is a case of 'de-extinction' by a trick of zoological nomenclature. These are all examples of supposed species extinctions but disappearances of moth populations are more easily documented. An example for Britain is the Feathered Ear (*Pachetra sagittigera*), which used to occur on the chalk downs of southeast England up until the early 1960s and is there no more.

SURVEYING MOTH DIVERSITY

Surveying moths is a popular pastime aided by a much wider range of field guides and websites than ever before. Many types of moth traps are available, or can be cheaply constructed with lightbulbs, notably ultraviolet (UV) ones, that are highly attractive to moths. Moths can be identified after dawn by carefully removing carton egg trays placed inside the trap. They like to rest in these, facilitating also the release of the trap contents. During the day, the trap should be kept shaded from the sun and its entrance lightly stuffed to prevent predators entering. Lists of accurate moth identities can be recorded either as species diversity (noting each species found) or true abundance data from counts of individuals observed in standardized traps. Such data are becoming increasingly useful to monitor local and natural population increases or, sadly, declines. In this way, rarer species and those that are first invasive or migratory to an area can be detected, much more so than day-flying species that rarely (if ever) come to light. In the future, benchtop or even handheld DNA sequencing technology may become widely available. Subsequent automated DNA analysis promises to boost objective ecological comparisons between traps and areas. Currently most such studies are still conducted by scientists, as it can be expensive involving a laboratory (see p. 140).

Another popular collecting method is sugaring, which consists of luring moths with highly attractive 'cocktails' of alcoholic drinks and sugars. This technique seems to be less susceptible to starry or moonlit nights and may help with monitoring species that are poorly attracted by lights. However, sugaring works best when there is less competition from natural food sources (such as sap runs or flowers) in the environment. For this reason, in cool temperate areas it works better from late summer to winter. Pirkka Utrio demonstrated that, amongst species flying late in the year, noctuids were more attracted than geometrids by complex mixtures that release many volatile compounds. Such an adaptation is linked to the need to gather highly energetically rewarding substances required to sustain the higher working temperature of the thoracic muscles in noctuids that fly as nights get colder.

Nevertheless, light-trapping is the main method used by researchers to sample moths. New light-source technology is being introduced as the baseline for finding moths in poorly explored remote areas. This requires either reflective sheets or easily portable traps with battery sources that can power a light for a whole night. There are increasing restrictions on toxic compounds such as mercury in UV lightbulbs and also on the capacity of lithium-ion batteries that may be carried on planes. Moth surveyors are thus regularly looking to adapt the latest light and battery technology for the best results. There is nothing more enjoyable than viewing the stunning range of forms and colours of moths that can be attracted to light, particularly in the tropics. With powerful LED light sources now available, many thousands of species can be found at a single site within several days, and using UV light, it is easy to gather new and rare species in the tropics. Recognizing that they are new is another matter. That requires a near comprehensive faunal and taxonomic knowledge. Moth taxonomists are still in demand.

BELOW Surveying moths in French Guiana using a white sheet illuminated by a lamp with strong UV emissions (invisible to us). The moths sit either on the sheet or the ground, where they can easily be inspected.

CHAPTER 8

Of moths and man

'LAY NOT UP FOR YOURSELVES TREASURES UPON EARTH, where moth and rust doth corrupt' the Bible exhorts. There is no doubt about it – moths have a bad name. Yet these delicate and beautiful creatures bring with them many positive attributes. They are themselves treasures on Earth. But before describing their usefulness, perhaps we should get their objectionable qualities out of the way.

THE DARK SIDE OF MOTHS

Moths do not always have the positive appeal to humans enjoyed by butterflies. Some people even find moths scary. Some moths, for example clothes moths, actively conflict with our activities (but so do some butterflies, notably cabbage whites). However, the relationship between man, moths and clothes is ambiguous. On the one hand, there are the reviled clothes moths with their reputation for destroying the contents of wardrobes. On the other, silk is one of our most prized fabrics and was a cornerstone commodity of international trade in the pre-industrial era, and is today a multi-billion dollar industry. We should also remember that moths were here long before us, and it is us that are now conflicting with them for resources on the planet.

UNWELCOME IN LARDERS, WAREHOUSES, PARKS AND FIELDS

Recently, clothes moths have been ravaging carpets and tapestries in museums and historic houses, as well as homes. The worst offenders are two main species, *Tineola bisselliella* and *Tinea pellionella*, which are very partial to fabrics containing wool. Nevertheless, certain once common clothes moths, like the Tapestry Moth (*Trichophaga tapetzella*) became rare after the introduction of central heating, perhaps because it prefers moister air. These moths are not limited to wool and will eat many sorts of furs, fabrics, food scraps and animal skin.

A number of other small moths are unwelcome in houses because they eat stored grains and flour. Among the most familiar are the European Grain Moth (*Nemapogon granella*), Indian Meal Moth (*Plodia interpunctella*), Meal Moth (*Pyralis farinalis*), Almond Moth (*Cadra cautella*), Raisin Moth (*C. figulilella*) and various *Ephestia* species, notably the Mediterranean Flour Moth (*E. kuehniella*). The larvae of these

OPPOSITE Women feeding silkworms on mulberry leaves – Chinese tempera painting on rice paper from the 19th century.

moths leave obvious silk webbing revealing their infestation. The webbing of flour moths has even been known to clog machinery in gristmills. These granivorous moths are generally able to feed on a wide variety of foodstuffs and have evolved to feed in generally dry and warm environments. This explains how they have easily adapted to human habitats, and became much more widespread thanks to our centrally heated houses. This distribution is also aided by transport in our storage vessels.

Another unwelcome impact of moths is seen in our gardens, parks, crop fields and plantations. Moths can even be perceived as aliens. In 2016 the British tabloids were busy trumpeting: 'Invasion of the Euromoths' and 'Mothball the EU'. This referred to the invasion of apparently billions of Diamond-back Moth (*Plutella xylostella*), whose caterpillars are addicts of cabbages and other crucifers. Larvae of several moth species are amongst the most serious crop pests in the world. Perhaps most notorious are the cutworms and armyworms (Noctuidae: Noctuinae, Heliothinae), cereal stem borers such as several Sesamiina noctuids and crambids (e.g. *Scirpophaga incertulas*, *Ostrinia nubilalis*) and the Codling Moth (*Cydia pomonella*; Tortricidae), which damages fruits such as apple and pear. Others are important defoliators of trees and every year, they devour tons of leaf mass from forests, orchards and ornamental trees. Just look at the damage to Horse Chestnut trees that has been caused in urban parks by its leaf miner, *Cameraria ohridella*, mentioned earlier (see p. 158), or the voluminous silken shelters of the Pine Processionary Moth (*Thaumetopoea pityocampa*) which dot thousands of hectares of Mediterranean pinewoods. In North America, the Gypsy Moth (*Lymantria dispar*) annually damages vast expanses of woodland. It was deliberately introduced there in the late 1860s for experiments on silk production by Étienne Léopold Trouvelot. Trouvelot's misstep was understandable given that both the Silkworm (*B. mori*) and Gypsy Moth were classified in the genus *Bombyx* at the time of these ill-fated efforts, but today the two moths are known to belong to distantly related families – a clear example of how costly bad taxonomy can be. This disastrous experiment in a species lacking natural controls resulted in forestry losses of millions of dollars per year. Tit for tat, the Fall Webworm (*Hyphantria cunea*) is an American 'gift' to the Old World. Defoliators of tropical forest plantations are many and include the Teak Moth (*Hyblaea puera*). Some species are injurious to orchards at the adult stage. These include the so-called fruit-piercing moths, above all the widespread (African and Indo-Australian) erebid species *Eudocima phalonia*, which with its strongly barbed proboscis, drills through the skin of fruits to feed on their juice, at the same time allowing infection with bacteria and fungi which cause the blemished fruits to rot.

APPOINTMENT WITH FEAR

It may be true that moths have a darker side, but some have a sinister symbolism. Perhaps that is also part of their broad cultural appeal. Throughout Europe the Death's-head Hawk-moth (*Acherontia atropos*) adorned with its skull-and-crossbones motif, has a reputation as an omen of death, or a curse like the plague. Actually this sinister looking moth is a robber of hives for their honey. Common names in other languages always stress the ominous death's head feature, such as Tête de Mort (French), Testa di Morto (Italian) and Totenkopf (German). *Acherontia atropos* featured in Bram Stoker's *Dracula*, Luis Buñuel's *Un Chien Andalou* and in paintings by William Holman Hunt (for example, *The Hireling Shepherd*) and Jan van Kessel's school. It inspired Edgar Allan Poe's *The Sphinx* and, with strong artistic license, also featured in paintings by Henry Fuseli (for example, *The Three Witches*). The moth gained infamy in *The Silence of the Lambs*. But between the book, the film and the poster, the moth became so hopelessly intermingled with other species (such as *Acherontia styx*, *Manduca sexta* and *Ascalapha odorata*) that disentangling them here would be an exercise in pedantry. A similar reputation has been gained by the Black Witch or Mariposa de la Muerte (*Ascalapha odorata*) in Latin America. In Madagascar, one of the few vernacular names for a moth is 'loloampaty', literally 'butterfly soul'. Interestingly, the ancient Greek word for Lepidoptera was 'Psyche', which meant both an individual 'butterfly' and the human soul, as following death, humans were thought to reincarnate as butterflies or moths. The loloampaty is related to the night-flying moth genus *Erebus* (a Greek name referring to the personification of darkness itself), and belongs to the genus *Cyligramma*, moths

LEFT The Lolompaty (*Cyligramma disturbans*) in the understory of the Madagascan rainforest. This moth is legendary in Malagasy culture as it can fly out of tombs where the ancestors are buried.

ABOVE The deadly caterpillar of a *Lonomia* (possibly *obliqua*), from Paraná (Brazil). This well armoured sickly green looking caterpillar can cause mortality in humans.

that frighten people as their dark wings sweep, bat-like, out of holes and tombs. Tombs are, to the Malagasy, the most sacred places from which their ancestors are annually disinterred and wrapped in silk shrouds made of another local moth, *Borocera cajani* (Lasiocampidae), in a 'dance with the dead'. DCL was once told by a Malagasy lady that, following her husband's death, an individual of *Cyligramma disturbans* magically appeared in her house. Filigreed on its dark wings is a pattern like a gold braid that was for her a sign of hope. A nice and detailed review of cultural lepidopterology is provided in Matthew Gandy's book *Moth*.

DEADLY MOTHS

On rare occasions moths are sinister, but can they be killers, though? Almost entirely not, but there are rare exceptions, one of which has become notorious, surprisingly only since the late 1960s. Caterpillars of the South American saturniid genus *Lonomia* are armed with spiny branched tubercles. Two species are reported in the medical literature *L. achelous* and *L. obliqua*, occurring in Venezuela and Brazil, respectively. Strangely beautiful but deadly, the caterpillars can cause internal haemorrhaging after prolonged contact, causing up to a few deaths per year, particularly in Brazil. Contact with *Lonomia* is known as 'lonomism'. A large number of individual spine-poison injections are required to cause death (fatalities from lonomism have been estimated at 2.5%), and it is mostly the very young or very old who are affected. One old lady died when a *Lonomia* caterpillar became trapped in her shoe.

MOTHS THAT ITCH AND STING

Several other saturniids (all from the generally toxic subfamily Hemileucinae) can be urticating, including the genera *Dirphia*, *Automeris*, *Leucanella*, and *Hylesia*. There are a number of terms in the medical literature for accidents caused by contact with such larvae, particularly 'erucism', and more generally, because hairs detaching from adult *Hylesia* moths can also cause reactions, 'lepidopterism'. The greyish hairy adults of this genus can send arrow-like bristles into the air causing a severe itching in humans and

RIGHT Pine processionary caterpillars (*Thaumetopoea pityocampa*) form a chain when searching a place to pupate. They should not be touched as the hairs can cause a serious rash.

their pets. French Guianans call this reaction 'papillonite' and Venezuelans 'Caripito itch'. Severe skin reactions are not only caused by Saturniidae, though. Among the tiger moths, another nasty South American moth is *Premolis semirufa*, which is quite polyphagous and does well in plantations of rubber trees (*Hevea brasiliensis*). Prolonged contact with the caterpillar's barbed hairs has caused 'Pararama associated phalangeal periarthritis' (to non medical people, joint inflammation) among Brazilian rubber tappers. Around the world, many caterpillars belonging to the families Erebidae (Arctiinae, Lymantriinae), Notodontidae (Thaumetopoeinae), Lasiocampidae, Limacodidae and Megalopygidae either cause skin irritation from their setae or intense nettle-like stinging (in the case of limacodids). The limacodid *Latoia vivida* in South Africa produces a sharp jabbing pain, becoming worst after 25 minutes and receding after two hours. These caterpillars inject histamines that are contained in a poison cell at the base of the tubercles. Their boldly brash and arty colours combined with pain is a remarkably effective deterrent to would-be predators. In parts of Madagascar, the vernacular name for caterpillar is 'sababaka'. Mention that word, and the locals instinctively scratch their arms.

Many Thaumetopoeinae (e.g. the Afrotropical genera *Anaphe* and *Hypsoides*) are urticating. In Europe and recently invading Britain, the Oak Processionary Moth (*Thaumetopoea processionea*) is at the front line of a suite of thaumetopoeine prominent moths that have rapidly expanded their ranges. The cocoons are removed from park trees by local councils, while the airborne barbed hairs of the caterpillars can cause rashes, not only in people but in pets, and when the latter get a caterpillar in the mouth it can cause a severe ulceration. These aerial setae are also a problem since they can cause sensitization. In Gotland a village population became sensitized to the release of hairs from *Thaumetopoea pinivora*. Also a problem are the larvae of tussock moths, Lymantriinae. To some people, encounters with larvae of the Brown-tail Moth (*Euproctis chrysorrhoea*) are rash ones. Municipal councils sometimes control outbreaks of Brown-tail Moth using flame throwers. 200,000 people in Japan in 1955 were affected by release of hairs from *Euproctis subflava* and there is a similar report for the same genus affecting around half a million people in Shanghai in 1981.

Perhaps the most venomous caterpillars in the USA are two flannel moth species in the genus *Megalopyge*. Megalopygid caterpillars are densely hairy, making them appear rather like a Skye terrier or Angora rabbit. Flannel moth caterpillars look like they have had bad comb-overs, but do not be tempted to stroke them! Below their long hairs are spines each equipped with a basal venom gland that charges them to their tips. This produces a pain that increases in intensity and can last for up to 12 hours while joints up and downstream of the site of envenomation can ache for two days. Totally extraordinary is the case of a bird that mimics such flannel moth caterpillars. The chicks of the Cinereous Mourner (*Laniocera hypopyrra*) in South America are bright orange and wildly hairy, their plumes bearing long white tips, resembling indeed the hairs of megalopygid caterpillars. An unattended chick that needs to protect itself will also imitate the creeping, peristaltic movements of such caterpillars.

TOP A megalopygid caterpillar from Peru.

ABOVE A Cinereous Mourner (*Laniocera hypopyrra*) chick mimics a megalopygid caterpillar both in appearance and behaviour. The white-tipped hairs of the caterpillar model are mimicked by the chick's feathers, as for its undulating movements, enhancing its resemblance to a stinging caterpillar when the parent is away.

VAMPIRE MOTHS AND EYE FREQUENTERS

Man is not normally attacked by vampire moths, but some Old World species of the genus *Calyptra* will draw blood from humans if given a chance. As we have seen in Chapter 3, several tropical moths are eye-frequenters. More than a dozen of these moths can be found feeding at night around a single buffalo eye, and there are also a few records of moths drinking tears from humans, as widely documented by Wilhelm Büttiker and Hans Bänziger. Some eye-frequenting moths in Africa spread *Trachoma* viruses among livestock. Both blood and tear-feeders have the potential to transmit bacteria and viruses to humans, but there are no documented cases.

THE BRIGHT SIDE

The injurious or sinister reputation of some moths is greatly outweighed by the ecological services they provide and the diverse uses that humans put moths and their products to, some of which can address environmental and technological problems.

MOTHS AS POLLINATORS

Moths include some critical pollinators, as well as many which supplement pollination services provided by other insects. A 2015 study by Callum Macgregor and collaborators found that 289 species in 75 plant families worldwide were partially or exclusively pollinated by moths, with 45 moth species known to pollinate orchids. Melanie Hahn and Carsten Brühl, in a 2016 study examining 227 moth-flower interactions in Europe and North America, found that 16 species of the pink and carnation family Caryophyllaceae also benefit significantly from moth pollination. The campion genus *Silene* is strongly dependent on *Hadena* (Noctuidae), even though larvae of *Hadena* later predate on maturing seeds. They have been termed 'parasitic pollinators' or, more generally, 'nursery pollinators'. Nottingham Catch-fly (*Silene nutans*) is one of the more famous moth-pollinated campions — evidently the moths (and caterpillars) are adept at avoiding its sticky stalks. The above studies highlighted the fact that most known pollination interactions involve Noctuidae and Sphingidae, whereas many other moth families are involved in pollination, indeed probably most ones that have a well-developed proboscis. These include Geometridae, for example the Speckled Yellow (*Pseudopanthera macularia*) which visits English bluebells. Many burnets and foresters (Zygaenidae) are also vital daytime pollinators of wild flowers. Moths tend to be overlooked though when talking about plant pollinators of gardens, but many garden ornamental plants other than pinks (*Dianthus*) such as *Delphinium*, *Aquilegia*, *Lonicera*, *Mirabilis* and lilacs (*Syringa*) are predominantly moth pollinated, as is *Buddleja*, notably at night the orange-ball species *B. globosa*.

Moths include some important crop pollinators. In Southeast Asia, oil palms are mostly pollinated by swarms of cosmet moths, *Pyroderces* (Cosmopterigidae). Some species of hummingbird hawkmoths (genus *Macroglossum*) and bee hawk moths (genus *Cephonodes*) supplement bee pollination in coffee plantations. In Madagascar, DCL twice observed swarms of *Ischnusia culiculina* (Zygaenidae)

around coffee flowers. The family Caricaceae is pollinated by a variety of moths. Papaya (*Carica papaya*), an important tropical fruit crop, is mainly sphingid pollinated. So is Ylang-ylang (*Cananga odorata*), plantations of which are found mainly in Comoros and Madagascar. Hawkmoths are the pollinators *par excellence* of sweet-smelling, long-tubed white flowers whose aroma comes into its own after dark. They powerfully disperse pollen over long distances, promoting plant outbreeding and genetic diversity. Tobacco plants (*Nicotiana*) are famously visited by long-tongued hawkmoths, especially the genera *Agrius* and *Manduca*. Ian Baldwin and colleagues found that *Nicotiana attenuata* plants lure *Manduca* moths with a benzylacetone aroma. Not only that, they regulate the dose of nicotine in the nectar to optimize the level of visitation by these specialist pollinators. That is partly because, although the tobacco plants need the moths to pollinate them, they would be harmed if their leaves are eaten by moth larvae. As a further means of protecting themselves against the hawkmoths depositing their eggs on the leaves, the plants can switch their flowers to morning opening. That would favour hummingbird pollination. A number of Solanaceae with sweet smelling flowers in the crepuscule (e.g. *Datura*, *Cestrum*, *Petunia*, *Atropa*), are also favoured for pollination by hawkmoths and so are many cacti whose flowers only open at night.

Many hawkmoth-pollinated plants like *Stephanotis floribunda*, *Gardenia*, Jasmine (*Jasminum*), Ylang-ylang, and Tuberose (*Polianthes tuberosa*) have such sweet, heady and overpowering fragrances, including terpenoid or benzenoid alcohols and esters, that they are used as important ingredients in the perfume industry, including more expensive and classic perfumes like Chanel N° 5. Exploring hawkmoth pollination preferences could even lead to a selection of new fragrances for the perfume industry.

Pyramidal Orchid (*Anacamptis pyramidalis*) is pollinated by many different moths, both nocturnal and diurnal. The stalks of its paired pollen masses, which lie on a

ABOVE LEFT A Speckled Yellow (*Pseudopanthera macularia*) pollinates a Common Bluebell (*Hyacinthoides non-scripta*) in an English woodland.

ABOVE RIGHT A Convolvulus Hawkmoth (*Agrius convolvuli*) pollinating Tobacco flowers in Italy.

sticky belt adhering to and wrapping around the proboscis of a visiting moth, are a special modification by this orchid for Lepidoptera pollination. Butterfly orchids (genus *Platanthera*) are all pollinated in fact by moths, whereas bog orchids (genus *Habenaria*) are pollinated especially by species of *Manduca* in addition to some long-tongued skipper butterflies. Orchids are renowned for indulging in deceptive pollination. Some are sufficiently good at mimicking floral scents typical in moth pollinated plants to get their pollen transferred from flower to flower, whilst often providing no reward such as nectar to the moths. Many orchids, like those of the genus *Aerangis* in East Africa studied by Steve Johnson and colleagues, provide a nectar reward to hawkmoths, but the most concentrated nectar is at the base of the nectar tube. The plants can thus economize on sugar production and provide the greatest reward to the longest tongued species. It is not only orchids that deceive moths. Frangipani (*Plumeria rubra*) yields no nectar reward, yet William Haber reported that it successfully deceives hawkmoths to visit its fragrant flowers. Female flowers of Papaya, which also lack nectar, provide another example of deceiving hawkmoths that anyway get nectar from the male flowers! One would think that was a risky strategy as hawkmoths can learn.

Some plants with specialized moth pollination risk extinction. This may be the case for the Vulcan Palm (*Brighamia insignis*; Campanulaceae) in Hawaii, a plant down to its last five populations, suggested to be pollinated by the Fabulous Green Sphinx (*Tinostoma smaragditis*), itself an endangered moth, rediscovered in 1998 on Kaua'i, although introduced *Manduca* species might serve as a backup to conservation efforts, currently relying on hand pollination.

The role of moths as pollinators is today severely threatened by artificial lights such as street lamps. We mentioned that Spindle Ermine moths which are light-flooded became more sedentary and less dispersive, thus by implication worse pollinators, after just a few generations. Moreover, a pollinator lured towards a light is not behaving as a pollinator and may just be eaten on the spot. Furthermore in 2017, Eva Knop and colleagues studying fields of thistles showed that they suffered significantly in seed set from the loss of moth pollination services in areas where there were streetlamps.

FERTILIZERS

When caterpillars are really abundant in forests (e.g. Winter Moth, Common Umber, Oak Tortrix, small ermines and Gypsy Moth), the fall of their droppings (frass) sounds just like raindrops and in extreme cases you might need an umbrella going for a walk! At such times, despite the obvious leaf skeletonization or defoliation of the forests, this is often part of a natural cycle that may not necessarily pose harm to the ecosystem. Rather, the outbreaks of most forest defoliators often occur when songbirds are nesting, when caterpillar abundance can result in greatly enhanced nesting success, including the production of an extra clutch of offspring. Furthermore, an intense and rapid recycling of the nutrients in digested leaves is happening during these outbreaks and the sheer mass of droppings may actually help to revitalize the forests, which can later flush with fresh leaves. Moth droppings are already being valued as an excellent fertilizer for pot plants. Now keen indoor gardeners can buy packets of caterpillar 'frass' as plant food. It is not only the frass that is rich in nutrients though. The huge global production of Silkworm (*Bombyx mori*) can be estimated as resulting in 90 billion pupae being boiled yearly for silk production (see Box, p.193). These pupae do not need to be wasted at all. They are eaten by man (see below), or they are used as the basis of feed for chickens, fish, and other livestock, but the chitinous cuticle as well as proteins of internal tissues also slowly release essential elements for plant growth such as nitrogen.

DINING ON MOTHS

Eating larvae has been termed 'anthropolarviphagy'. In many parts of the world, moth caterpillars provide a vital source of protein for humans. François Malaisse and Paul Latham recorded over 100 moth species of 11 different families as being eaten in Africa, even as a staple in particular seasons. In southern Africa, Mopane Worm (*Gonimbrasia belina* and other saturniid species) form the basis of a tens of millions dollar industry annually. In rural Botswana, up to 40% of protein requirements can be met by consumption of these caterpillars. In Burkina Faso, Charlotte Payne recounts how local women venture out well before dawn to collect the larvae of the Pallid Emperor Moth (*Cirina forda*) from Shea trees. This is likely an age-old practice for hominids in the region. A fossil pupa closely resembling this genus was unearthed in the palaeontological site Laetoli (Tanzania), dating back more than three million years. In Mexico, 17 species of Lepidoptera have been identified as edible, mostly moths belonging to 11 different families. In Mexico, agave worms are popularly eaten, famously the worm ('gusano') found at the bottom of bottles of Mezcal, a distillate of various species of agave, but not Blue Agave (*Agave tequilina*), which is made into Tequila, lacking the 'worm' (actually a caterpillar). Such larvae are also sometimes encapsulated in lollipops. Gusanos are rarely white agave worms – the caterpillars of skipper butterflies (Hesperiidae, tribe Megathymini) that can fetch $250 per kilogram when exported. Usually, cheap moth substitutes are used particularly Red Agave Worm (*Comadia redtenbacheri*; Cossidae), or even a weevil. Eating caterpillars has a long tradition in Mexico. Larvae of the spectacular Neotropical looper moth, *Pantherodes pardalaria*, have been traded there since antiquity. The Bamboo Worm (*Omphisa fuscidentalis*; Crambidae) is very popular in Thailand. A villager can collect as much as 50 kg in a day from bamboo stems, and such 'fried bamboo worms' contain about 25% protein and 50% fat.

There does not appear to be a word for 'eating pupae'. Yet in Madagascar, recently formed pupae are much savoured over larvae – a strong cultural difference from Africans. Malagasy people would not normally want to eat a bowl of brightly coloured caterpillars. In southern Madagascar, the pupae of the Madagascan Emperor (*Antherina suraka*) are much valued for their protein. Humans have evolved to eat insect protein. The nutritional advantages are considerable since protein plus lipid content often exceed 55%. Silkworm pupae, extensively consumed in the Far East, are known to be rich also in vitamin B12, α-linolenic acid and minerals. The giant witchetty grubs of Australia (families Hepialidae and Cossidae) have doubtless been sought out

ABOVE Stylized moth paintings (reminiscent of Thyrididae) made by bushmen in Raiders 1 Cave in the Brandberg Massif, Namibia. This shows that moths have not only been important as food but have long inspired art in humans.

ABOVE A pot of Mopane Worm (*Gonimbrasia belina*). Mopane Worm larvae are vital as a protein source in poor African countries at some times of the year.

BELOW Bamboo 'pot' containing the fat juicy caterpillars of the Bamboo Worm (*Omphisa fuscidentalis*).

by aborigines from the stems in which they feed for tens of thousands of years. There is archaeological evidence from caves in Australia dating back a thousand years or more that aborigines ground aestivating Bogong Moth (*Agrotis infusa*) into a fine and delicious paste using stone mortars.

Moths even have tremendous potential directly as feed for livestock. It is far better for the planet and for the health of livestock and pets to feed them insect protein than to destroy the Amazon for soya production. Moreover, it takes far more energy and hundreds of times more water to produce equivalent amounts of beef than mealworms. Some companies are now mass rearing the Greater Wax Moth (*Galleria mellonella*) specifically for this purpose. An advantage of mass-producing moths for food and feed or even energy bars (as already being done from mealworm and cricket flour) is that the industrial process removes any potential microbial contaminants that might be a health concern when eating caterpillars and pupae neat. Also, insect protein has zero risk of mad cow disease and lower risk for *Salmonella*. Greater Wax Moth larvae are widely sold also as fishing bait, but in Australia, the caterpillars of Bardee Grub (*Trictena atripalpis*; Hepialidae) are also used.

MOTHS AS INFUSIONS AND DRUGS

Some of the most expensive and coveted teas in the world are made out of the percolated frass (droppings) of several moths in the Orient, notably *Bombyx mori* in Taiwan. Such teas are often reputed to have restorative qualities. In the mountainous areas of Guangxi, Anhui and Guizhou Provinces of China, the tea Chóng Chá, dark and fragrant, is derived from the frass of *Hydrillodes morosa* (Erebidae) that was fed on *Platycarya strobilacea* (Juglandaceae) or *Litsea coreana* (Lauraceae). This tea is also made from the frass of the Black Rice Worm or Tea Tabby (*Aglossa dimidiata*), reared on *L. coreana* or the apple tree *Malus seiboldii*. Supposedly, Chóng Chá fights heatstroke, soothes digestion and relieves intestinal bleeding.

There is another indirect way that moths are sought after for medicinal purposes. On the Tibetan Plateau, as well as in Himalayan parts of India and Bhutan to 4,650 m elevation, locals go in search of subterranean dead caterpillars, hoping to find 'ergots'. These are the finger-like projections of Dōng Chóng Xià Cǎo (*Ophiocordyceps sinensis*), a fungus that grows from the dead bodies of swift moth caterpillars such as the Himalayan Bat Moth (*Thitarodes armoricanus*), and other related species. *Ophiocordyceps sinensis* is regarded as one of the most expensive mushroom in the world since its over 20 active compounds are used to treat a panoply of diseases, particularly of a respiratory or immune nature, but also ailments of the kidney, liver and heart. Tibetan herdsman 1,500 years ago first noticed that their yaks became energetic after the animals ate them. These fungi are hugely important in Chinese culture. They were used as far back as the Tang dynasty (618–907 AD) and even regarded as an elixir of youth. The Chinese beauty Yang Kue Fei (719–756 AD) was particularly keen on them.

BIOMIMETICS

Moths inspire technology. This field of reverse engineering from nature is known as biomimetics. One of the most impressive nanotechnological advances triggered by looking at moths has been the development of anti-reflective arrays based on the surface of moth compound eyes, which have regular narrowly tapered or conical nanostructures that help reduce incident light scattering (see p. 16). Such nanotechnology allows the design of mobile phone screens that can be read in bright light and allows office windows to be made much more efficient in transmitting light. Antireflection coatings can cut down reflection to 0.23% or less. Such coatings can also make solar cells up to 6% more efficient. Finally, a hypersensitive camera was built by NASA influenced by the configuration of moth eyes. Nanotechnology is increasingly being inspired by the physical structure and properties of Lepidoptera scales. New research is coming out all the time, for example investigating the possibility of using the scales of the Sunset Moth (*Chrysiridia rhipheus*) to generate a sensor of pH conditions (acidity), as a 'biological physical dye'. The automotive industry has been investigating new paints which have reflective glows mimicking those on a moth's eye.

(a)   (b)

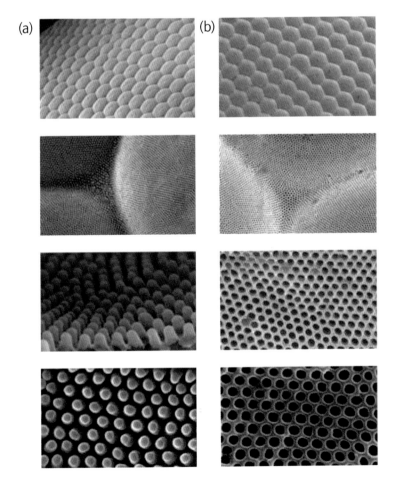

ABOVE A Greater Wax Moth (*Galleria mellonella*) caterpillar munches plastic. That these caterpillars can complete their development on plastic has inspired research into organisms capable of degrading a major scourge of our times.

LEFT Highly magnified conical surface structures of the compound eye of an Atlas moth (*Attacus atlas*) (a). When such nanostructures are replicated via a polyether mold (b), they can be used to create water-repellent, anti-reflective coatings. Such reverse engineering is known as biomimetics.

BIOTECHNOLOGY

In an extraordinary discovery in 2017, a technician noticed holes in a plastic bag. A team led by Federica Bertocchini realised that the Greater Wax Moth (*Galleria mellonella*) has the ability to feed on and break down plastics. It now appears that Greater Wax Moth larvae host gut bacteria that help them break down polyethylene and even polypropylene. Another team had already noticed in 2014 that larvae of Indian Meal Moth (*Plodia interpunctella*) can bite holes in plastic bags and can degrade polyethylene films, evidence that this ability is not limited to moths that eat wax. These discoveries, along with silk biotechnology (see below) are of major potential significance for the waste industry and are galvanising the culture of bacteria which have such plastic-degrading properties for use on an industrial scale.

THE THREAD THAT JOINED EAST AND WEST

The ancient trade route that links the West with the Orient is known as the Silk Road, after the highly desirable silk fabric that was its major trade item. It is fair to say that in addition to trade, this route catalysed the exchange of knowledge and culture since the Silk Road's beginnings around 200 BC. Before tradesmen began travelling this route, China had managed to keep the secret of the silkworm for four to five millennia – so long, in fact, that the domesticated Silkworm adult no longer bears much resemblance to its nearest wild relative, the Wild Silk Moth (*Bombyx mandarina*). The glossiness and softness of silk generated by the caterpillars of the long inbred Silkworm (*Bombyx mori*), has long made it a highly desirable natural fabric. In fact, other silk moths which were used in the West, as well as Africa and Madagascar, did not produce such a coveted fabric as the Silkworm. China fiercely guarded its industrial secret until, by one historical account, monks sent out by Emperor of the East, Justinian I (482–565 AD), succeeded to smuggle and bring Silkworm 'eggs' (or possibly cocoons?) to Constantinople, hiding them in canes. Intellectual property theft was happening thousands of years ago. There is evidence that the bombycid *Rondotia menciana* was used thousands of years ago in China, too.

In a few other places, local people made do with their local silk moths. In Oaxaca province in Mexico, Cortez brought the Silkworm from Spain as far back as 1523. He successfully introduced this new species to a native weaving tradition based on the communal silk nests of the butterfly *Eucheira socialis*. The Silkworm did not reach the USA until the early 1600s, when silk enthusiast King James I tried, and failed, to establish a Silkworm industry in Jamestown, Virginia. In parts of Europe – before silkworm technology and mulberry cultivation finally seeped out of China – the main silk provider was the eggar moth *Pachypasa otus*. It is thought that the diaphanous silk of the Greek island of Kos, referred to as Coan silk by Aristotle and Pliny, derived from this species. Although the Indus valley used wild silk moths back to at least 2450 BC, it is the diversity of sericulture from 'wild' silk moths in the Oriental region that is most significant. Tussar (= Tasar, tussah, tussore) silk from *Antheraea pernyi* and about five related species including Japanese Oak Silkmoth (*A. yamamai*), which is strong and elastic, has the longest history of utilisation, dating back 3,500 years in

THE VALUE OF SILKWORMS

Only two invertebrates can claim the accolade of giving rise to multi-million dollar industries: *Bombyx mori* and *Apis mellifera*. Silk has been a prized and sought after commodity for at least 4,500 years. For hundreds if not thousands of years it was China's most important export. Though commodities such as amber, glass, spices and tea were also traded in the Middle Ages between East and West, the dominance of silk is underscored by the name given to the principal trading route between East and West in the pre-industrial era: the Silk Road. Silk is still an important commodity today. Global production of silk averaged approximately 175,000 metric tonnes per year between 2013 and 2017, of which almost 150,000 metric tonnes per annum was produced in China. Almost 1 million workers are employed in the silk sector in China and the silk industry provides employment to 7.9 million people in India. The value of the global silk market is difficult to assess but one recent research report projects that the value of the silk market will reach $16.94 billion by 2021. Given that the value of honey production in the United States in 2017 was only $318 million, the value to man of the silkworm comfortably exceeds that of the honeybee.

China, and the industry was exported to India 600 years ago. In India, 'Tasar' is also produced from *A. mylitta* and *A. paphia*. Muga, glossy and golden silk for royalty, is produced from *Antheraea assamensis*, while the Ailanthus Silkmoth (*Samia cynthia*) is still used to produce the darker, coarser Eri silk, warm for winter and cool for summer wear.

Moth silks are not just used for clothing. The Camphor Silkworm (*Eriogyna pyretorum*) produces silk tough enough for use as fishing lines and in surgery, but since the 1910s, it has gone out of commercial production. Silk is made of two proteins, sericin and fibroin, and it is the latter that is especially important in medicine. Silk from Silkworm, but especially the very strong Tussar silk, can be used as an organic scaffold for tissue regeneration, including bone, to manufacture inserted targets for drug delivery, and in wound healing. Other applications of silk are endless. Silkworms can spin super-silk by from mulberry leaves if they are sprayed with carbon nanotubes or graphene, which can provide reinforced clothing and conductive fabrics. To reduce the growing mountains of non-recyclable waste, natural silk can be manufactured into biodegradable coffee cups, silk food wrap, dental floss and other commonly discarded items. Also of great potential, wild silk can help local people generate income at the same time as enhancing the natural environment. Today, local community-based initiatives in the tropics are trying to diversify wild silk use and sericulture. In the last few years in Madagascar, one NGO has been successful in breeding the local silk moth *Ceranchia apollina* (p. 23). It has also succeeded in spinning the giant communal processionary moth cocoons (*Hypsoides* sp.) into yarn but their urticating setae need first to be removed. Such local sericulture is highly sustainable, encouraging the planting of native forest plant species while the breeding avoids wild harvesting.

BIOLOGICAL CONTROL

The most famous case of moths being used in biological control (i.e. the fight against undesired organisms with natural enemies or competitors), is that of the Cactus Moth (*Cactoblastis cactorum*). This was imported from Argentina into

RIGHT Larvae of the aptly named *Cactobastis cactorum* devour prickly pears (species of *Opuntia*). This pyralid moth was introduced from Argentina into Australia and South Africa to control this cactus where the latter is invasive.

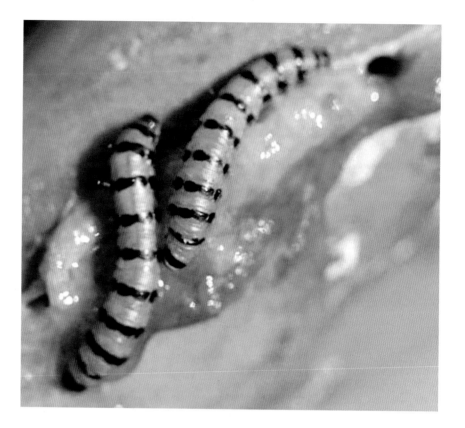

Eastern Australia from 1914 for eradicating prickly pears (*Opuntia* species) that were originally introduced in the late 18th century for live fencing, but eventually smothered vast land expanses. So successful was this control that the moth was introduced elsewhere. However, in South Africa, its introduction diminished the amount of spineless *Opuntia* available as cattle fodder. Much worse, its introduction into the Caribbean nations has proved to be ill-advised as the moth island-hopped its way to the North American mainland, where it is proving to be an ominous threat to native, including globally imperiled species of *Opuntia*. Worse still, the moth is continuing to expand its range northward and westward through Gulf States, and now poses a threat to Mexico. Not only are planted prickly pears an important food source, but Mexico is rich in native *Opuntia*. Considerable money and effort must now be spent annually to limit the spread of this hero turned villain. The beautiful Cinnabar Moth (*Tyria jacobaeae*) is widely used to control Tansy Ragwort (*Jacobaea vulgaris*), in the western USA alongside the Ragwort Flea Beetle. A more recent proposal is to use moths as biological control in the war against drugs, to target plantations of Coca (*Erythroxylum coca*) in Colombia, using mass release via aircraft of the native Coca Tussock Moth (*Eloria noyesi*; Erebidae), but the measure has not yet been adopted by the government. Moths are also used to control weeds. In Hawaii, entomologists released Madagascan Fireweed Moth (*Galtara extensa*) to control the increasingly cosmopolitan weed Madagascar Ragwort (*Senecio madagascariensis*), whose seeds had possibly been inadvertently introduced in mulch from Australia.

KEEPING NATURE IN CHECK

Vastly more important than their use in plant control, moth caterpillars worldwide perform a tremendous ecological service. They keep in check the luxuriant growth of leaves, allowing plant diversity space to flourish, cleaning up waste vegetable matter, and even digest horns and hooves in the African savannah, in the case of the incredible Horn Borer (*Ceratophaga vastella*).

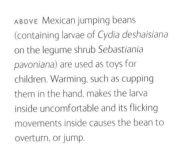

MOTHS FOR RECREATION AND INSPIRATION

Moths can also be used as toys for children. The most famous example is probably the Mexican Jumping Bean Moth (*Cydia deshaisiana*), the pod abodes of which are shed mainly from the tree *Sebastiana pavoniana*, and they actually saltate as a result of the larvae jerking when warmed (the usual trick is to warm up the beans in the hand). It appears the caterpillars can move the beans when they are exposed to heat. In the right conditions, these tortricids will eventually pupate and emerge through an exit hole in the bean.

Moths are aesthetically very appealing, exhibiting striking beauty, wild and mesmerising colours, and sometimes ghostly or just bizarre appearance. Their likeness has always been represented in art. Nowadays they are widely used in advertisements, posters and films. Their images are among the most shared on photographic forums and their memes often go viral on social media. Their colours and patterns provide inspiration for some of the latest fashion designs. There is an undeniable cuteness also to caterpillars. Breeding moths is a fantastic recreational activity for both children and adults that can not only teach about life cycle and biology, but provide great pleasure in observing natural transformations.

In many countries, comprehensive guides to the moth fauna are now online, so that it is easy to identify moths you observe in your back garden by digital means. Moths have great potential in citizen science. Moth-watching, and especially moth photography, are taking off as a new great craze after bird, butterfly, and dragonfly watching. Get a moth trap and help discover your local biodiversity thus providing a baseline for monitoring changes in the environment.

MOTHS IN A CHANGING WORLD

A few moth losses we know about, but most we do not. These are the silent, unsung majority, extinguished before they become known to science. The natural enemies of moths are detailed elsewhere, but sadly their most important foe is humans. Some have proposed a new geological era called the Anthropocene (starting about

ABOVE Mexican jumping beans (containing larvae of *Cydia deshaisiana* on the legume shrub *Sebastiania pavoniana*) are used as toys for children. Warming, such as cupping them in the hand, makes the larva inside uncomfortable and its flicking movements inside causes the bean to overturn, or jump.

BELOW An anthropomorphic Death's-head Hawk-moth (or a 'mothmorphic' lady...) features in one of the trade cards used by Company Liebig in the 1890s to popularize its meat extracts.

1950). We are now living in this era, defined by human influence, in which the natural rate of extinctions is increased by perhaps three orders of magnitude, and whatever its relative level, it far exceeds the generation of new species.

The major anthropogenic threats to moths are direct habitat loss and deforestation (e.g. by logging and mining), and agricultural intensification with its tendency towards monocultures that are maintained by pesticides and fertilizers. All this is linked to a creeping urbanization, complete with its spheres of light pollution. The ecological footprint of our modern way of life is almost unimaginable, with three quarters of terrestrial environments globally already impacted. Added to these is now the most insidious threat, climate change. The inexorable rise in atmospheric greenhouse gases, notably carbon dioxide (from 274 parts per million around 1900 to 415 ppm today) as a by-product of the industrial revolution, has caused a dramatic rise in temperatures (global warming). This trend shows every sign of accelerating, putting more energy and water into the atmospheric system and so extreme climate events, which can certainly kill populations of moths, are becoming more and more frequent. In some parts of the world, there is a growing body of data that human activities have reduced moth populations drastically. Alarming headlines now abound in the press, with phrases like 'insect apocalypse' or 'insectageddon', and writers recall past days when driving yielded a regular debris of insect splats on windscreens, clogged grills and headlights reflected a snowstorm of moths in the summer.

Habitat fragmentation is becoming an impediment to the dispersal of moths. Even if some moth species eke out an existence in minute areas, not enough habitat and genetic diversity of populations may remain to sustain many in the long term. Some moths could literally be the 'flying' dead. The reduction in viable primary habitats affects islands more than continents. Species native to the smallest islands, geographic or ecological, are the most vulnerable. Transport networks in particular bring with them a massive ramifying edge effect through their original habitat. Providing for biodiversity corridors is important but the effect of fragmentation is exacerbated by the inability of some species to disperse between surviving habitat fragments. For example, species of Micropterigidae which live in places like Japan and New Zealand, where populations are very localized, are highly prone to extinction. On Madagascar, where numerous micropterigid species have not yet been formally described, each one seems restricted to a particular biotope in a single mountain range. Because so little has been documented about such low dispersive moth groups, we have practically no idea how many have been eradicated in the last century by the destruction of vast swathes of tropical forest.

The human role in extinctions is shown starkly by detailed data on the overall decline of moth populations. Studies by Butterfly Conservation (a charity which also focuses on moths!) conducted over several decades in the UK show that between 1968 and 2007, 37% of larger moth species declined in abundance on average by an amplitude of 28%. Elevational transects (i.e. samples taken at a series of stations going up a mountain slope), such as those by Jeremy Holloway on Mt Kinabalu in Borneo which were replicated after 42 years, showed that the elevational ranges of

102 moth species moved upslope over this period by an average of 67 m (220 ft). Such elevational shifts are mirrored by latitudinal movements as the planet warms, with some species expanding not only into higher elevations but higher latitudes, at the expense of the local populations. In fact, it is likely that the species adapted to the lowest temperatures (e.g. at mountaintops) eventually will have nowhere left to expand into. A suite of moth species, notably ones invading from other countries, instead take advantage of anthropogenic change, but the overall result may be that such species will tend to dominate local faunas. At least 1% of Europe's moths are invasive. The threats mentioned above, when combined, provide a perverse negative synergy affecting moths.

Extinctions and declines may seem a wistful place to end this book. Yet increasingly we realise the discrepancy between what we know about moths and what we do not. Around 140,000 species of moths are known to science, but some estimates would put the true global diversity to maybe five times this figure. It is quite hard to discover a new species of moth in a well-known temperate area like the British Isles, but stunning numbers of unknown species can be found in the tropics. There is thus a tremendous challenge to sample dwindling habitats and discover more about, not only the unknown species, but the many unstudied life histories of moths. It is almost futile to design a mitigation plan to conserve or rescue a declining species without first understanding its specific ecological requirements, not least its larval hosts.

In summary, while moths sometimes have a sinister reputation, or for a few people their manic fluttering can cause fear, actually those species that can cause us harm or pain or ravage our crops are really few. In fact, the myriad of moths are among the foremost herbivores, keeping plant growth in check, and they have a vital role in ecosystems, not just by pollinating flowers but by providing a link between plants and other organisms in food webs, transferring nutrients and energy from plants into tasty morsels for other animals. For thousands of years, silk moths have clothed and adorned humans and indeed, trade in their silk was part of the opening of economies and the cross fertilization of cultures. Their early stages provide human cultures with a vital fallback of protein in times of famine. Moths inspire poets and novelists as well as artists. They fill museums with their beautifully prepared and long lasting exoskeletons. Such collections are both a record of the past and a scientific treasure trove for the future. Moths incessantly inspire scientific, biomedical and nanotechnology research and they provide a source of nagging questions including still the oldest and most basic, why is a moth lured to a flame?

Moths are among the most appealing yet mysterious organisms on the planet. A stanza of Thomas Hardy about a couple staring into a candlelight sounds apt if issued to the unfortunate moth that turned into a cinder in front of them:

"What are you still, still thinking, He asked in vague surmise,
That you stare at the wick unblinking, With those great lost luminous eyes?"

Appendix

Higher taxonomy of moths, showing placement of all families to superfamily (see p.4 for the branching patterns of the latter) with the approximate numbers of genera and species in each group, based on a 2011 count, with minor modifications. 'Clades' are more or less inclusive groupings which comprise all superfamilies and clades that are listed further down.

The butterfly superfamily (Papilionoidea) is not highlighted in colour. Throughout the book we refer to 'micromoths', a very loose term of convenience including all Lepidoptera that are not in the clade Macroheterocera, some of which are 'macromoths'. See map p.156 as a key to principal biogeographic regions.

Superfamily/family	Common or other names	Approx. # genera	Approx. # species	Distribution
Order Lepidoptera (extant groups; 42 superfamilies; 129 families)		15353	157829	
Superfamily Micropterigoidea	micropterigoids	24	265	Worldwide
Family Micropterigidae	micropterigids or jaw moths	24	265	Worldwide
Superfamily Agathiphagoidea	agathiphagoids	1	2	Australasia (E-Australia, Melanesia, Polynesia)
Family Agathiphagidae	agathiphagids or Kauri moths (*Agathiphaga*)	1	2	Australasia (E-Australia, Melanesia, Polynesia)
Superfamily Heterobathmioidea	heterobathmioids	1	10	Southern South America
Family Heterobathmiidae	heterobathmiids or Southern Beech moths (*Heterobathmia*)	1	10	Southern South America
Clade Glossata or Tongue moths (all below, 39 superfamilies)		15326	157551	Worldwide
Superfamily Eriocranioidea	eriocranioids	7	30	Holarctic
Family Eriocraniidae	eriocraniids or purples	7	30	Holarctic
Superfamily Lophocoronoidea	lophocoronoids	1	6	Australasia (Australia)
Family Lophocoronidae	lophocoronids (*Lophocorona*)	1	6	Australasia (Australia)
Superfamily Mnesarchaeoidea	mnesarchaeoids	1	7	New Zealand
Family Mnesarchaeidae	mnesarchaeids or New Zealand primitive moths (*Mnesarchaea*)	1	7	New Zealand
Superfamily Hepialoidea	hepialoids	69	629	Worldwide
Family Hepialidae	hepialids or swifts	69	629	Worldwide
Superfamily Neopseustoidea	neopseustoids	4	13	Southern South America and Himalayas-East Asia
Family Neopseustidae	neopseustids	4	13	Southern South America and Himalayas-East Asia
Family Aenigmatineidae	aenigmatineid or Kangaroo Island Moth (*Aenigmatinea glatzella*)	1	1	Australasia (SE-Australia)
Family Acanthopteroctetidae	acanthopteroctetids	2	8	Worldwide (W-USA, Europe, S Africa, Andes)
Clade Heteroneura (all below, 34 superfamilies)		15244	156866	Worldwide
Superfamily Nepticuloidea	nepticuloids	29	1054	Worldwide
Family Nepticulidae	nepticulids or pigmy moths	22	862	Worldwide
Family Opostegidae	opostegids or white eyecap moths	7	194	Worldwide
Superfamily Andesianoidea	andesianoids	1	3	Southern South America (Andes)
Family Andesianidae	Andean endemic moths (*Andesiana*)	1	3	Southern South America (Andes)
Superfamily Adeloidea	adeloids	58	927	Worldwide
Family Cecidosidae	cecidosids or gall moths	6	16	South America, South Africa and New Zealand
Family Prodoxidae	prodoxids or yucca moths	8	97	Nearctic, E-Palearctic (Russia and Japan)
Family Tridentaformidae	tridentaformid (*Tridentaforma fuscoleuca*)	1	1	North America (California)
Family Incurvariidae	incurvariids	12	51	Worldwide, especially Palearctic
Family Heliozelidae	heliozelids	12	124	Worldwide
Family Adelidae	adelids or longhorns	5	294	Worldwide
Superfamily Tischerioidea	tischerioids	3	112	Worldwide
Family Tischeriidae	tischeriids or trumpet leaf-miners	3	112	Worldwide
Superfamily Palaephatoidea	palaephatoids	7	57	Southern South America and Australasia (Australia)
Family Palaephatidae	palaephatids	7	57	Southern South America and Australasia (Australia)

Clade Ditrysia (all below, 29 superfamilies)		**15147**	**154714**	Worldwide
Superfamily Tineoidea	**tineoids**	**574**	**3719**	Worldwide
Family Meessiidae	meessiids	2	90	Holarctic (mainly)
Family Psychidae	psychids or bagworms	210	1200	Worldwide
Family Eriocottidae	eriocottids	6	80	Worldwide
Family Dryadaulidae	dryadaulids or dancing moths (*Dryadaula*)	1	46	Worldwide
Family Tineidae	tineids or clothes moths and allies	355	2303	Worldwide
Superfamily Gracillarioidea	**gracillarioids**	**117**	**2205**	Worldwide
Family Roeslerstammiidae	roeslerstammiids or double-eye moths	13	53	Worldwide
Family Bucculatricidae	bucculatricids	4	297	Worldwide
Family Gracillariidae	gracillariids	100	1855	Worldwide
Superfamily Yponomeutoidea	**yponomeutoids**	**230**	**1735**	Worldwide
Family Yponomeutidae	yponomeutids incl. ermine moths	95	352	Worldwide
Family Argyresthiidae	argyresthiids (*Argyresthia*)	1	157	Worldwide
Family Plutellidae	plutellids incl. diamond-back moth	48	150	Worldwide
Family Tonzidae	tonzids (*Tonza*)	1	3	Old World tropics
Family Glyphipterigidae	glyphipterigids incl. sedge moths	29	535	Worldwide
Family Ypsolophidae	ypsolophids	7	155	Worldwide
Family Attevidae	attevids (*Atteva*)	1	52	Worldwide
Family Praydidae	praydids	3	45	Worldwide
Family Heliodinidae	heliodinids	13	69	Worldwide
Family Bedelliidae	bedelliids (*Bedellia*)	1	16	Worldwide
Family Lyonetiidae	lyonetiids	32	204	Worldwide
Clade Apoditrysia (all below, 26 superfamilies)		**14225**	**147054**	Worldwide
Superfamily unassigned	**—**	**3**	**4**	Palearctic, Chile
Family Millieriidae	millieriids	3	4	Palearctic, North America (Florida), S South America (Chile)
Superfamily Douglasioidea	**douglasioids**	**31**	**134**	Worldwide
Family Douglasiidae	douglasiids	2	29	Palearctic
Superfamily Simaethistoidea	**simaethistoids**	**2**	**4**	Himalayas, Australia
Family Simaethistidae	simaethistids	2	4	Himalayas, Australia
Superfamily Alucitoidea	**alucitoids**	**21**	**234**	Worldwide
Family Tineodidae	tineodids	12	19	Oriental, Australasian
Family Alucitidae	alucitids or many-plumed moths	9	215	Worldwide
Superfamily Pterophoroidea	**pterophoroids**	**93**	**1307**	Worldwide
Family Pterophoridae	pterophorids or plume moths	93	1307	Worldwide
Superfamily Carposinoidea	**carposinoids**	**28**	**326**	Worldwide
Family Copromorphidae	copromorphids	9	43	Worldwide except Palearctic
Family Carposinidae	carposinids or fruitworm moths	19	283	Worldwide except NW-Palearctic
Superfamily Schreckensteinioidea	**schreckensteinioids**	**2**	**8**	Holarctic, Neotropics, Oriental
Family Schreckensteiniidae	schreckensteiniids	2	8	Holarctic, Neotropics, Oriental
Superfamily Epermenioidea	**epermenioids**	**10**	**126**	Worldwide
Family Epermeniidae	epermeniids or fringe-tufted moths	10	126	Worldwide
Superfamily Urodoidea	**urodoids**	**3**	**65**	Worldwide, mostly Neotropics
Family Urodidae	urodids or false burnet moths	3	65	Worldwide, mostly Neotropics
Family Ustyurtiidae	ustyurtiids (*Ustyurtia*)	1	2	Palaearctic (Kazakhstan)
Superfamily Immoidea	**immoids**	**6**	**245**	Pantropical
Family Immidae	immids	6	245	Pantropical
Superfamily Choreutoidea	**choreutoids**	**18**	**406**	Worldwide
Family Choreutidae	choreutids or metalmark moths	18	406	Worldwide
Superfamily Galacticoidea	**galacticoids**	**3**	**19**	Old World tropics; North America (introduced)
Family Galacticidae	galacticids incl. mimosa webworm	3	19	Old World tropics; North America (introduced)
Superfamily Tortricoidea	**tortricoids**	**1043**	**9757**	Worldwide
Family Tortricidae	tortricids or tortrix moths	1043	9757	Worldwide

Superfamily Cossoidea	cossoids	359	2745	Worldwide
Family Brachodidae	brachodids or little bear moths incl. pseudocossines	14	137	Worldwide
Family Cossidae	cossids or goat, leopard and carpenter moths	134	857	Worldwide
Family Dudgeonidae	dudgeonids	6	56	Old World tropics
Family Metarbelidae	metarbelids	16	183	Afrotropics
Family Ratardidae	ratardids	3	10	SE-Asia
Family Castniidae	castniids incl. sun and butterfly moths	34	113	Worldwide, mainly Neotropics
Family Sesiidae	sesiids or clearwings	152	1389	Worldwide
Superfamily Zygaenoidea	**zygaenoids**	**541**	**3284**	Worldwide
Family Epipyropidae	epipyropids	9	32	Worldwide mainly tropical
Family Cyclotornidae	cyclotornids (*Cyclotorna*)	1	5	Australia
Family Heterogynidae	heterogynids (*Heterogynis*)	1	10	Western Mediterranean
Family Lacturidae	lacturids	8	120	Pantropical
Family Phaudidae	phaudids	3	15	SE-Asia
Family Dalceridae	dalcerids or glass moths	11	80	Neotropics
Family Limacodidae	limacodids or slug and cup moths	298	1660	Worldwide
Family Megalopygidae	megalopygids or flannel moths	23	232	Neotropics
Family Aididae	aidids	2	6	Neotropics
Family Somabrachyidae	somabrachyids	4	8	Afrotropics
Family Himantopteridae	himantopterids or long-tailed burnet moths	11	80	Oriental and Afrotropics
Family Zygaenidae	zygaenids incl. burnet and forester moths	170	1036	Worldwide
Clade Obtectomera (all below, 13 super families)		**12065**	**128394**	Worldwide
Superfamily Gelechioidea	**gelechioids or curved-horn moths**	**1608**	**20652**	Worldwide
Family Lypusidae	lypusids	3	21	Palearctic
Family Chimabachidae	chimabachids	2	6	Palearctic, Nearctic (introduced)
Family Stenomatidae	stenomatids	30	1200	Largely Neotropical
Family Depressariidae	depressariids	92	802	Worldwide
Family Peleopodidae	peleopodids	14	68	Worldwide
Family Ethmiidae	ethmiids	5	278	Worldwide
Family Oecophoridae	oecophorids or concealer moths	313	3216	Worldwide
Family Lecithoceridae	lecithocerids or long-horned moths	94	1158	S-Palearctic and Old World tropics
Family Xyloryctidae	xyloryctids	60	524	Indo-Australasia (mainly), Afrotropics, Polynesia
Family Autostichidae	autostichids	72	638	Worldwide
Family Elachistidae	elachistids or grass-miner moths	161	3201	Worldwide
Family Batrachedridae	batrachedrids	10	99	Worldwide
Family Pterolonchidae	pterolonchids	2	8	Palearctic and South Africa, Nearctic (introduced)
Family Syringopaidae	syringopaid (*Syringopais temperatella*)	1	1	Palearctic (Mediterranean), Far East to India
Family Coelopoetidae	coelopoetids (*Coelopoeta*)	1	3	Nearctic (North America)
Family Momphidae	momphids	6	115	Holarctic (mainly), Neotropics, New Zealand, Madagascar
Family Coleophoridae	coleophorids or case-bearer moths	5	1386	Worldwide, mainly Nearctic
Family Blastobasidae	blastobasids	26	370	Worldwide
Family Scythrididae	scythridids or flower moths	30	668	Worldwide
Family Stathmopodidae	stathmopodids	44	408	Worldwide, mainly tropics
Family Cosmopterigidae	cosmopterigids or cosmet moths	135	1777	Worldwide
Family Gelechiidae	gelechiids or twirler moths	500	4700	Worldwide
Family Epimarptidae	epimarptids (*Epimarpta*)	1	4	W-India
Family Schistonoeidae	schistonoeid (*Schistonoea fulvidella*)	1	1	Pantropical (?origin West Indies)

		No. genera	No. species	Distribution
Superfamily Papilionoidea (7 families) butterflies incl. skippers and New World night butterflies		1806	18539	Worldwide
Superfamily Calliduloidea	**calliduloids**	**7**	**49**	Madagascar, E-Palearctic and Indo-Australia
Family Callidulidae	callidulids or Old World butterfly moths	7	49	Madagascar, E-Palearctic and Indo-Australia
Superfamily Hyblaeoidea	**hyblaeoids**	**3**	**19**	Pantropical (mainly Old World)
Family Hyblaeidae	hyblaeids or teak moths	2	18	Pantropical (mainly Old World)
Family Prodidactidae	prodidactid (*Prodidactis mystica*)	1	1	Afrotropics (Southern Africa)
Superfamily Whalleyanoidea	**whalleyanoids**	**1**	**2**	Madagascar
Family Whalleyanidae	whalleyanids (*Whalleyana*)	1	2	Madagascar
Superfamily Thyridoidea	**thyridoids**	**93**	**940**	Worldwide
Family Thyrididae	thyridids or leaf moths	93	940	Worldwide
Superfamily Pyraloidea	**pyraloids**	**2075**	**15576**	Worldwide
Family Pyralidae	pyralids or snout moths	1055	5921	Worldwide
Family Crambidae	crambids	1020	9655	Worldwide
Superfamily Mimallonoidea	**mimallonoids**	**27**	**194**	Neotropics
Family Mimallonidae	mimallonids or sack-bearer moths	27	194	Neotropics
Clade Macroheterocera (all below,	**5 superfamilies)**	**6445**	**72423**	Worldwide
Superfamily Drepanoidea	**drepanoids**	**126**	**672**	Worldwide
Family Cimeliidae	cimeliids or gold moths	2	6	Mediterranean
Family Doidae	doids	2	6	North America to Northern South America (incl Caribbean)
Family Drepanidae	drepanids or hook-tips	122	660	Worldwide
Superfamily Bombycoidea	**bombycoids**	**497**	**4357**	Worldwide
Family Apatelodidae	apatelodids or American silkworm moths	10	145	Neotropics up to North America
Family Eupterotidae	eupterotids or monkey moths	53	339	Old World tropics and Australia
Family Brahmaeidae	brahmaeids or owl and brahmin moths and *Lemonia*	7	65	Palearctic, SE-Asia and Afrotropics
Family Phiditiidae	phiditiids	4	23	Neotropics
Family Anthelidae	anthelids or Australian lappet moths	9	94	Australia
Family Carthaeidae	carthaeid or Dryandra Moth (*Carthaea saturnioides*)	1	1	Australia
Family Endromidae	endromids, incl. Kentish Glory	12	58	Palearctic and Oriental
Family Bombycidae	bombycids or silkworms	26	185	Worldwide
Family Saturniidae	saturniids or silkmoths and emperor moths	169	2004	Worldwide
Family Sphingidae	sphingids, hawkmoths or hornworms	206	1443	Worldwide
Superfamily Lasiocampoidea	**lasiocampids**	**224**	**1948**	Worldwide
Family Lasiocampidae	lasiocampids or eggars	224	1948	Worldwide
Superfamily Geometroidea	**geometroids**	**2116**	**23770**	Worldwide
Family Epicopeiidae	epicopeiids or oriental swallowtail moths	9	20	SE-Asia and E-Palearctic (Himalayas to Japan)
Family Sematuridae	sematurids	6	40	New World. Southern Africa
Family Uraniidae	uraniids	97	713	Worldwide (mainly tropics)
Family Pseudobistonidae	pseudobistonids	2	2	SE-Asia (India to Himalayas, China, Thailand)
Family Geometridae	geometrids, loopers or inchworms	2002	22995	Worldwide
Superfamily Noctuoidea	**noctuoids**	**3483**	**41677**	Worldwide
Family Oenosandridae	oenosandrids	4	8	Australia
Family Notodontidae	prominents	450	3200	Worldwide
Family Erebidae	erebids incl. semiloopers, snouts, tiger and tussock moths	1741	24470	Worldwide
Family Euteliidae	euteliids	28	522	Worldwide
Family Nolidae	nolids	186	1738	Worldwide
Family Noctuidae	noctuids or owlets	1074	11739	Worldwide
TOTAL MOTH SPECIES APPROXIMATELY		13550	139000	
TOTAL LEPIDOPTERA SPECIES APPROXIMATELY		15350	158000	

Further information

FURTHER READING

Bourgogne, J. 1951. *Ordre des Lépidoptères.* In: Grassé, P.-P. (editor), Traité de Zoologie, 10 (1). Masson et Cie, Paris.

Common, I.F.B. 1990. *Moths of Australia.* Melbourne University Press, Victoria.

Covell, C.V. (Jr.) 2005. *A Field Guide to Moths of Eastern North America.* Virginia Museum of Natural History, Martinsville.

Dett, A. (editor). 2018. *Moths of Costa Rica's Rainforest.* Benteli, Salenstein.

Dugdale, J.S. 1988. *Lepidoptera - annotated catalogue, and keys to family-group taxa.* Fauna of New Zealand, 14. Science Information Publishing Centre, Wellington.

Ford, E.B. 1972. *Moths.* 3rd edn. New Naturalist, 30. Collins, London.

Forster, W. & Wohlfahrt, T.A. 1954. *Die Schmetterlinge Mitteleuropas, I. Biologie der Schmetterlinge.* Franckh, Stuttgart.

Gandy, M. 2016. *Moth.* Reaktion Books, London.

Grimaldi, D. & Engel, M.S. 2005. *Evolution of the Insects.* Cambridge, New York.

Heppner, J.B. 1998. *Classification of Lepidoptera.* Holarctic Lepidoptera, 5 (1). Association for Tropical Lepidoptera, Gainesville.

Holloway, J.D., Kibby, G. & Peggie, D. 2001. *The Families of Malesian Moths and Butterflies.* Brill, Leiden.

Holloway, J.D., Bradley, J.D. & Carter, D.J. 1987. *CIE Guides to Insects of Importance to Man, 1. Lepidoptera.* CAB International Institute of Entomology & British Museum Natural History, Wallingford.

Hooper, J. 2003. *Of Moths and Men: Intrigue, Tragedy and the Peppered Moth.* Norton, London.

Inoue, H. (editor). 1982. *Moths of Japan.* 2 vols. Kodansha & Co., Tokyo.

James, G.D. 2017. *The Book of Caterpillars: A life-size guide to six hundred species from around the world.* The Ivy Press, London.

Komai, F., Yoshiyasu, Y., Nasu, Y. & Saito, T. (editors). 2011. *A Guide to the Lepidoptera of Japan.* Tokai University Press, Kanagawa.

Krenn, H.W. 2010. Feeding mechanisms of adult Lepidoptera: structure, function, and evolution of the mouthparts. *Annual Review of Entomology,* 55: 307-327.

Kristensen, N.P. (editor). 1999 & 2003. *Lepidoptera, Moths and Butterflies.* Handbuch der Zoologie, IV. Arthropoda: Insecta, 35. 2 vols. W. de Gruyter, Berlin

Majerus, M.E.N. 2002. *Moths.* New Naturalist, 90. Collins, London.

Mayr, E. 1942. *Systematics and the Origin of Species.* Columbia University Press, New York.

Pinhey, E.C.G. 1975. *Moths of Southern Africa.* Tafelberg, Cape Town.

Portier, P. 1949. *La Biologie des Lépidoptères.* Lechevalier, Paris.

Powell, J.A. & Opler, P.A. 2009. *Moths of Western North America.* University of California Press, Berkeley.

Quicke, D.L.J. 2017. *Mimicry, Crypsis, Masquerade and other Adaptive Resemblances.* Wiley-Blackwell, Oxford.

Robinson, G.S., Tuck, K.R. & Shaffer, M.P. 1994. *A Field Guide to the Smaller Moths of South-East Asia.* The Natural History Museum & Malaysian Nature Society, Kuala Lumpur.

Scoble, M. 1992. *The Lepidoptera: Form, Function and Diversity.* The Natural History Museum & Oxford University Press, Oxford.

Stehr, F.W. (editor). 1987. *Immature Insects, 1.* Kendall-Hunt Publishing Co., Dubuque.

Sterling, P. & Parsons, M. 2012. *Field Guide to the Micromoths of Great Britain and Ireland.* British Wildlife Publishing, Gillingham.

Wagner, D.L. 2005. *Caterpillars of Eastern North America: a Guide to Identification and Natural History.* Princeton University Press, Princeton.

Waring, P. & Townsend, M. 2003. *Field Guide to the Moths of Great Britain and Ireland.* British Wildlife Publishing, Hook.

Zborowski, P. & Edwards, T. 2007. *A Guide to Australian Moths.* CSIRO, Canberra.

A MORE COMPREHENSIVE BIBLIOGRAPHY CAN BE FOUND AT

Natural History Museum Open Repository https://nhm. openrepository.com/.

MAIN BOOK SERIES

Die Gross-Schmetterlinge der Erde (A. Seitz, editor); Microlepidoptera of Europe; Microlepidoptera Palaearctica; Monographs of Australian Lepidoptera; Moths of Borneo (by J.D. Holloway); Moths of America North of Mexico; Moths of Thailand; Noctuidae Europaeae; The Geometrid Moths of Europe

SOCIETIES

The Lepidopterists' Society https://www.lepsoc.org/; The Lepidopterological Society of Japan http://lepi-jp.org/; Lepidopterists' Society of Africa https://www.metamorphosis.org.za/; Societas Europaea Lepidopterologica [SEL] http://www.soceurlep.eu/

INTERNET RESOURCES

Afromoths http://www.afromoths.net/; AnimalBase http://www.animalbase.org/; Atlas of Living Australia [ALA] https://www.ala.org.au/; Barcode of Life Database [BOLD] http://www.boldsystems.org; Biodiversity Heritage Library [BHL] https://www.biodiversitylibrary.org/; Bibliothèque Nationale de France [BnF] https://gallica.bnf.fr/accueil/en/content/accueil-en?mode=desktop; Butterflies and Moths of the World: Generic Names and their Type-species https://www.nhm.ac.uk/our-science/data/butmoth/search/; Caterpillars, pupae, butterflies & moths [Costa Rica, Guanacaste]; http://janzen.sas.upenn.edu/caterpillars/database.lasso; Digital Moths of Japan http://www.jpmoth.org/; Encyclopedia of Life [EOL] https://www.eol.org/;

Funet http://www.nic.funet.fi/pub/sci/bio/life/insecta/lepidoptera/index.html; Global Biodiversity Information Facility [GBIF] https://www.gbif.org/en/; Google Books https://books.google.com/; Google Scholar https://scholar.google.com/; HOSTS - a Database of the World's Lepidopteran Hostplants https://www.nhm.ac.uk/our-science/data/hostplants/; iNaturalist https://www.inaturalist.org/; Internet Archive https://archive.org/; Lepiforum http://www.lepiforum.de/; Lepindex https://www.nhm.ac.uk/our-science/data/lepindex/; Moths and Butterflies of Europe and North Africa, https://www.leps.it/; Moths of Borneo http://www.mothsofborneo.com/; Moths Photographers Group https://mothphotographersgroup.msstate.edu; Natural History Museum Data Portal https://data.nhm.ac.uk/; Naturkundliches Informationssystem [NKIS] http://www.nkis.info/nkis/; Papillons de Poitou-Charentes http://www.papillon-poitou-charentes.org/; Papua Insects Foundation https://www.papua-insects.nl/; Tree of Life http://tolweb.org/tree/; UKmoths https://ukmoths.org.uk/; Zobodat Literature https://www.zobodat.at/publikation_series.php

Beginner's guide to species, systematics, taxonomy and nomenclature

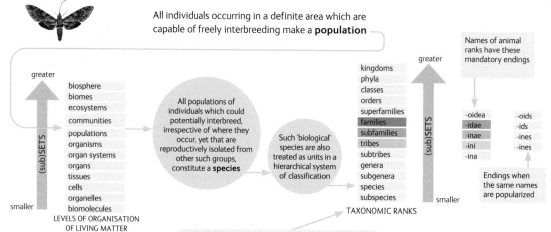

Glossary

Note: many adjectives in the text starting in lower case pertain to a 'taxonomic' group, e.g. agaristine to the noctuid subfamily Agaristinae. Such words are not in the glossary but pp.138–139 and the Appendix, pp.198–201. may be checked to resolve how they are anglicized. For taxonomic groups down to family level, see Appendix. Here we list only the more important and less obvious terms, or those mentioned more than once. Some terms, identical to those used for other animal groups, refer to different structures.

adaptive radiation evolutionary splitting into multiple species coupled with diversification in body features allowing exploitation of wide array of ecological opportunities or niches in a geographic area

aedeagus the phallus

aestivation period of dormancy over the summer (counterpart to **hibernation**)

allopatric geographically separate

amplexiform wing coupling system relying on basal expansion of the hindwing overlapping broadly with forewing

anagenesis the process where a single evolutionary lineage changes over time without branching into more lineages

androconia (sing. **androconium**) specialized scent-releasing scales of male

angiosperm flowering plant, whose ovules are enclosed within an ovary (this structure turning into a fruit)

aposematism use of conspicuous signals to advertise distastefulness

apparency exhibition of resources that a host (e.g. plant) may offer to exploiters (e.g. herbivores)

aptery/apterism wingless condition

ballooning aerial dispersal on strands of silk

Batesian mimicry resemblance of toxic or otherwise well-defended 'model' species by 'mimics' actually quite palatable to predators

brachyptery (adj. **brachypterous**) overall wing reduction

bursa copulatrix purse-like copulatory structure of female genitalia – see fig. p.36

calyx the outer whorl of parts including sepals in a flower, which protects the developing petals

canopy the upper tree layer in a forest

chaetosema (pl. **chaetosemata**) raised areas on the head containing arrays of sensory bristles

chaetotaxy analysis of the pattern of distribution of sensory setae on the body

chitin (adj. **chitinous**) polysaccharide compound that is key constituent of the insect exoskeleton or cuticle

chromosomes greatly coiled thread-like structures in cells carrying the genetic material

cladogenesis evolutionary splitting of a lineage (e.g. species) into daughter lineages

cline gradual divergence of features along an ecological or geographic gradient

cloaca common outlet of reproductive duct(s) and intestine

commensalism loose ecological association where one partner benefits and the other is at any rate not harmed

compound eye(s) principal eye(s) of adult insects, composed of multiple ommatidia

coremata (sing. **corema**) inflatable tubular sacs that are expanded from male abdomen for disseminating volatile compounds

cornuti thorn-like structures on vesica of male genitalia – see fig. p.36

countershading contrast in colour intensity shading counteracting the revealing effect of cast light

coxa most basal leg segment that articulates with or is fused to the thorax

crepuscular dusk-active

crochet series of minute hooks on the planta of larval prolegs ensuring firm grip

cyanogenic glucosides cyanide-binding compounds, toxic to predators when they release hydrogen cyanide

cytokinin plant hormone involved in breakdown of pigments in leaf

diapause dormancy during the life cycle that is hormonally-induced

dimorphism (adj. **dimorphic**) having two forms; the most common example being sexual dimorphism

disruptive breaking up a typical or expected appearance of a potential prey item

distal further away from body core

DNA barcode sequence of a fixed, up to 658 base pair long nucleotide fragment of the **Cytochrome Oxidase I** (**COI**) gene, often used like a DNA fingerprint to distinguish (identify) species – see box p.140.

dorsal upper part or topside (counterpart of ventral); pertaining to dorsum or 'back'

ductus seminalis duct for transfer of sperm from bursa copulatrix to spermatheca in female genitalia

ecdysis skin change (moulting) during post-embryonic development, typically of a larva

ecdysone moulting hormone inducing ecdysis (e.g. leading mature larva to pupate)

eclosion emergence of a caterpillar from an egg ('hatching') or of an adult from a pupa

ecological niche the multidimensional ecological setting in which a particular species can live

ectoparasite external parasite

elevation vertical height above sea level

endemic exclusively occurring in a defined geographic area

epiphysis cleaning brush of foreleg – see fig. p.33

exoskeleton (= **cuticle**) external protective and structural covering of an arthropod

exuvia (pl. **exuviae**) discarded cuticle of a caterpillar or pupa

femur main leg segment joined to tibia – see fig. p.33

flagellum shaft of antenna

flocculent secretion that spontaneously turns into filamentous spray when emitted

founder(s) the original colonist(s) that following dispersal successfully establish(es) in a new area

frass droppings of a caterpillar

frenulum bristle from the hindwing that forms part of a forewing/hindwing linking structure

frons front of the head – see fig. p.44

fusiform spindle-shaped

galea (pl. **galeae**) outer lobe of the maxilla, which originates the paired halves of adult proboscis – see fig. p.44

genotype the genetic constitution of an organism

gnathos ventral part of 10th abdominal segment in male genitalia, a paired or fused structure under the uncus

gymnosperm seed-bearing vascular plant, e.g. conifer, with ovules not enclosed in an ovary (so no fruits are produced)

gynandromorph mixed sex individual – see p.55

haemolymph insect blood

hair pencil bundled tuft or brush of hair-like androconial scales

head capsule exoskeleton of head, which is shed as a cap at each moult

herbarium a collection of pressed plant specimens

heteroneurous forewing and hindwing venation different

holoptic greatly enlarged compound eyes, joining at the midline

homoneurous forewing and hindwing venation similar

hybrid zone area where ranges of closely related species overlap and hybrids between them are produced

hypermetamorphosis dramatic change of morphology or appearance between instars

hypopharynx fleshy, spiny lobe of caterpillar mouthparts used for swallowing – see fig. p.44

invasive species a species that arrives and locally establishes from an area outside its place of detection

isotope ratio measurement by mass spectrometry of variants of a particular chemical element differing in numbers of neutrons

jugum small forewing lobe in primitive moths with limited wing linking function

juvenile hormone maintains insect in a larval stage, inhibiting adult characters

juxta shield-like part of male genitalia – see fig. p.36

kairomone plant volatile compound that a parasitoid wasp uses to locate its host

keratin structural protein lining skin of four-legged vertebrates, making up nails, claws, hooves, fur etc.

labial palp multiply jointed outer part of labium (usually reduced in primitive Lepidoptera) – see fig. p.44

labium ventralmost part of mouthparts deriving from fusion of an ancestral pair of secondary maxillae – see fig. p.44

labrum dorsalmost part of mouthparts overhanging mandibles (both structures reduced in most adult Lepidoptera) – see fig. p.44

lek assemblage of a single sex, typically males, to attract the opposite sex and compete for mates

lineage an evolutionary unit, not necessarily a species, at any rank of the classification hierarchy – see bottom of opposite page

maxillae paired part of mouthparts for cutting and manipulating food including galea – see fig. p.44

maxillary palp multiply jointed outer part of maxilla (often reduced in lees primitive adult Lepidoptera) – see fig. p.44

mesothorax second (middle) segment of thorax

metamorphosis (**complete metamorphosis**) process undergone by organisms that completely reshape their body during development

metathorax third (posterior) segment of thorax

micropyle small perforated plate in egg for entry of sperm

mimicry ring an assemblage of different species sharing an apparently convergent signal (e.g. colour pattern) in the same general area or environment

monomorphism having one form, thus no phenotypic variants (see also phenotype and polymorphism)

monophagous feeding on a single plant species (term sometimes applied when feeding on a single plant genus)

Müllerian mimicry shared resemblance among toxic species (co-mimics) to maximise benefit against predation

myrmecophily ecological relationships of organisms regularly associated with ants

myrmecoxene infrequent visitor to ant's nests that functions as an occasional 'guest'

necrosis cell death; may visibly affect tissues by exhibiting dark patches

ocellus (pl. **ocelli**) simple eyes bearing only one lens, generally adjacent to compound eye(s); applied also to fake eyespots

oligophagous feeding on a few plant species or genera

ommatidium (pl. **ommatidia**) single unit(s) of a compound eye

ostium bursae genital opening used by female for mating

oviduct egg duct of female genitalia

ovipore egg laying opening of female genitalia – see fig. p. 36

ovipositor often extrusible egg laying apparatus of female genitalia

palps (**palpi**) appendages of insect mouthparts (see labial palp, maxillary palp) – see fig. p.44

papillae anales pair of setose lobes flanking ovipore and anus at the end of the female abdomen

parapatric geographically contiguous or minimally overlapping (as pertains to a range)

parasitoid organism injecting its egg into a host and ultimately killing it after the parasitoid larvae grow

parthenogenesis virgin birth

patagia (sing. **patagium**) pair of flaps on prothorax

pecten a comb of bristles on the antenna

pectinate comb-like, e.g. bearing side branches like the tines of a comb, as in feathery antennae

pedicel basal segment of the antenna between scape, where this articulates with the head, and flagellum

peripatric speciation evolutionary process by which species originate by the geographic budding off of small population(s) from a parent species range

pharate a fully formed stage concealed under the covering of a preceding stage (e.g. an adult moth under pupal cuticle)

phenotype the outward appearance of an organism

pheromone chemical released by an individual as a signal that affects the behaviour of another individual of the same species

phoresy (adj. **phoretic**) a relationship where one organism uses another as its carrier

photoperiodism developmental response of phases in the life cycle to changing daylengths

phototactic movement in response to light – either positive (towards) or negative (away from) light

phylogenetic pattern or 'tree' (in short, **phylogeny**) a diagrammatic reconstruction of evolutionary relationships among biological entities, e.g. between species, as an inference of ancestor-descendent relationships

planta soft extrusible pad on a proleg of a caterpillar

plastron gill an air bubble formed by a water-repellent coating allowing breathing underwater

pleurae lateral parts of a body segment

polymorphic occurring in many distinct forms or variants (see also monomorphism, dimorphism)

polyphagous feeding on many plant species or genera

polyphenism multiple discrete variants of phenotype resulting from a single genotype under differing environmental conditions

preadaptation a pre-existing configuration of genes or other traits allowing exploitation of a new ecological opportunity

pre-imaginal early stage (egg, larva, caterpillar), counterpart to imaginal (from **imago**, adult)

prepupa motionless final stage of a caterpillar during which radical changes happen before entering the pupal stage

pretarsal pertaining to the terminal segment of the leg (**pretarsus**) bearing claws – see fig. p.33

prolegs fleshy pairs of supplementary legs ('false legs') on larval abdominal segments used to grip substrate and walk

prothorax first (anterior) segment of the thorax

pyrrolizidine alkaloid (PA) a type of nitrogenous, toxic plant compound

radiation a group of species sharing a common ancestor that have evolutionarily diversified (radiated) in a particular geographic area (see also adaptive radiation)

range the actual geographic area where a species (or more inclusive groupings such as genera, families etc.) is distributed

reflex bleeding production of droplets of repellent substance (usually haemolymph) to deter potential predators

reproductive isolation an essential result of speciation, when species lose reproductive compatibility with other species

retinaculum catch on the forewing for receiving the hindwing frenulum

rhabdom sensory element of an ommatidium – see fig. p.16

scape basal segment of the antenna articulating with head

scale typically hollow outgrowth of insect cuticle clothing body parts including wings, where they are usually flattened in shape

sclerite a discrete hardened plate of the exoskeleton

sclerotin cross-linked protein providing structural reinforcement to cuticle, via a process called tanning

sclerotized hardened with sclerotin

seasonal polyphenism variation determined by different environmental conditions experienced during development by different generations of a same species

secondary contact pattern and process where previously contiguous populations that have subsequently split into two or more geographically separated areas (i.e., are allopatric) regain contact; at this stage differences can be sharpened up

secondary setae setae devoid of sensory function that appear after the first moulting in a caterpillar; often long and hair-like

semivoltine taking more than a year to complete life cycle (see also voltinism)

sensillum (pl. **sensilla**) microscopic cuticular structures (e.g. hair-like or as small cavities) of a sensory nature, such as those on a single branch of an antenna

seta (pl. **setae**; adj. **setose**) small hair-like projections (bristles) on the cuticle

sibling equivalent member of a brood, e.g. a brother or sister

sibling species species that are practically identical morphologically but may differ in certain key biological traits (e.g. pheromones) – see box p.167

signa sclerotized spines or plates inside the bursa copulatrix of the female genitalia

sister species a pair of species which are evolutionarily most closely related to one another and derived from a common ancestor – see box p.167

spatulate broadened at tip, spatula-like

speciation evolutionary process by which species are produced

spermatheca small sac used for temporary storage of sperm by female

spermatophore sac containing sperm transferred as viscous fluid from male which hardens in female

spinneret spindle-shaped protrusion of the labium of caterpillar mouthparts used for everting silk – see fig. p.44

spiracle opening or valve within the exoskeleton for breathing

stemmata simple eyes (typically 5-6) in a larva

stenopterous with wings too narrow to sustain flight

sternal from ventral side (**sternum**) of a body segment

stigma pollen receiving part of flower (female counterpart of stamen)

stridulation sound production

structural colour physically produced colour resulting from optical interaction of light with cuticular structures, not pigment

stylet needle-like or scalpel-like mouthpart

symbiosis living together for mutual benefit

sympatric co-occurring in a same area

tapetum reflective layer at back of eye

tarsus terminal leg segment – see fig. p.33

tegulae scaled triangular pieces like shoulder pads

tegumen hood-like dorsal part of 9th abdominal segment in male genitalia – see fig. p.36

tergal from the dorsal side (**tergum**) of a body segment

terpenoid defensive compound in resinous plants such as conifers

thorax middle of three main subdivisions in insect body (dividing head and abdomen), the one bearing wings and walking legs

tibia leg segment

tracheae (adj. **tracheal**) a complex system of tubes that internally ramify from the spiracles for respiration – see fig. p.35

triturating basket a kind of crop in the mouth for grinding up pollen grains or spores

trochanter usually highly reduced leg segment between coxa and femur

true legs in a caterpillar, the legs of the three thoracic segments (see also prolegs; 'false legs')

trunk the section posterior to the head along the body

tymbals sound emitting organs comprising membranes that resonate when buckled

tympanate possessing hearing organs containing a membrane (**tympanum**) at opening of acoustic chamber that transmits its vibrations to a tiny stretch receptor organ

type specimen specimen serving as a fundamental anchor point of scientific naming or nomenclature (e.g. holotype)

uncus often hook-like dorsal part of 10th abdominal segment in male genitalia – see fig. p.36

urticating irritating; in moths usually via sharp or dehiscent setae, sometimes which have embedded toxins

valve clasping part of 9th abdominal segment in male genitalia (paired) – see fig. p.36

ventral lower part or underside (counterpart to dorsal); pertaining to venter or 'belly'

vertex top of the head

vesica internal membranous sac of phallus, everted during copulation – see fig. p.36

vicariance process of splitting of geographic ranges by their fragmentation or by emergence of geographic gaps

vinculum U-like or V-like ventral part of 9th abdominal segment in male genitalia – see fig. p.36

voltinism number of generations of a species over a year (univoltine, bivoltine, multivoltine, etc.).

Index

Picture credits

pp.4 and 35 ©Sandra Doyle; p.7, 46(bottom), 105(top), 119(top) ©Andreas Kay; p.8, 18, 24(bottom), 25(top), 63, 101, 114(top.left), 122, 158, 183 ©David C. Lees; p.10, 13(top, right), 44(top), 45(right), 77(top), 109, 112, 113(bottom) 114(top, right), 115, 116(top.left and bottom), 117(bottom), 125(bottom), 172(top) ©John Horstman; p.11 © Jan Fischer Rasmussen; p. 12 (bottom, left), p.14 (right), 64©Steve Gschmeissner/Science Photo Library; p.13 (bottom, left) ©Ilona Loser p. 14 (left) © Rudolf Bryner; p.16 (bottom)©JB Wheatley; p.19 ©Steve Irvine; p.21(top) ©Dennis Kunkel Microscopy/Science Photo Library; p.21 (bottom), 26(bottom), 65(bottom)), 95, 169, 170(top), 172(bottom) ©Patrick Clement; p.22 ©DJTaylor/shutterstock.com; p.23(bottom), 49(bottom) ©Florian Teodor/shutterstock.com; p.24(top) ©Marek Mis/ Science Photo Library; p.25(bottom) ©guentermanaus/ shutterstock.com; p.26(top) © Nevit Dilmen [CC BY-SA 3.0 (https://creativecommons.org/licenses/by-sa/3.0)]; p.27 Francisco Welter-Schultes [CC0]; p.31, 66(bottom), 96, 185(top) ©Ingo Arndt/naturepl.com; p.32 ©Josef Stemeseder/ shutterstock.com; p.33(bottom) ©Friedmar Graf; p.37 ©Ignác Richter; p.38 ©Kenji Nishida; p.40©Cosmin Manci/ shutterstock.com; p.41 ©Jussi Murtosaari/naturepl.com; p.42 ©Dr Neal G. Smith; p.43 ©Carol Snow Milne; p.45(left)/ Alpsdake [CC BY-SA 3.0 (https://creativecommons.org/licenses/ by-sa/3.0)]; p.46(top) ©Eerika Schulz; p.47 ©Leroy Simon/ Visuals Unlimited, inc./Science Photo Library; p.48 © Mike Dunn, roadsendnaturalist.com; p.49(top and right) ©Chekaramit/ shutterstock.com; p. 50 ©Erick Greene; p.51 (left) ©Pan Xunbin/Science Photo Library; p.51 (right), 98(bottom)© Gyorgy Csoka, Hungary Forest Research Institute, Bugwood.org; p.52(left) ©Dafinka/shutterstock.com; (left); (right) ©entomart; ©Eric Warrant; p.54 ©Henrik Larsson/shutterstock.com; p.58 ©Nathan Yuen; p.59 ©Daniel Rubinoff; p.61 Lawrence E. Reeves; p.62 ©Arthur Anker; p.65 (top)© J K Lindsey; p.66 (top) © Muhammad Mahdi Karim/Wikimedia Commons; p.67, 78 ©Melvyn Yeo/Science Photo Library; p.48 © Takashi Komatsu; p.68 (bottom) Dejean, A. et al. A cuckoo-like parasitic moth leads African weaver ant colonies to their ruin. Sci. Rep. 6, 23778; doi: 10.1038/srep23778 (2016); p. 70 ©Stephen Barlow; p.71(top) ©Liz Milla; p.71(bottom) ©Atsushi Kawakita; p.73© Lutz Thilo Wasserthal; p.74 ©Roland Hilgartner; p.76 ©John Lampkin; p.77 (bottom) ©Piotr Naskrecki/Minden/naturepl.com; p.80(top) ©RachenStocker/shutterstock.com; p.80 (bottom, left) ©Steve Ogden; (bottom, right) ©Breck P. Kent.shutterstock.

com; p.81 (top) ©Peter Buchner; p.81 (bottom) ©Nicky Bay; p.83 (top) ©Sarah2/shutterstock.com; p.83 (bottom) © Neil Hewett, Daintree Rainforest - Cooper Creek Wilderness; p.84(top) ©Jean and Fred/Flickr; p.85 ©Wolfgang Wagner; p.88(top) ©Axel Hausmann, SNSB, Zoologische Staatssammlung Muenchen; p.88(bottom) ©William E. Conner et al. PNAS 2000;97:26:14406-14411; p.89(top) Modified after Phelan & Baker (1990. Journal of Insect Behavior, Vol. 3 (3), fig. 1); p.89(bottom) ©Michel Candel; p.90(top) ©Ian Adams; (bottom) ©Brian Peers; p.92 ©Colin Varndell/Science Photo Library; p.93 © Ko, Doo-Hyun & R. Tumbleston, John & Henderson, Kevin & E. Euliss, Larken & M. DeSimone, Joseph & Lopez, Rene & Samulski, Ed. (2011). Biomimetic microlens array with antireflective "moth-eye" surface. Soft Matter. 7. 6404-6407. 10.1039/ C1SM05302G.; p.94 ©Charlie Streets; p.98(top), 150, 157 Alex Hyde/naturepl.com; p.99 ©IDmitry Fch/shutterstock.com; p.100 ©Mark R. Shaw; p.102 ©Dr Morley Read/Science Photo Library; p.103(top) ©Dante Fenolio/Science Photo Library; p.103 (bottom) ©Richard Hoyer; p.104 © Brian Kunkel, University of Delaware, Bugwood.org; p.105 (middle), 111 ©Pete Oxford/ Minden/naturepl.com; p.105(bottom) Matt Edmonds; p.106 ©Nick Green; p.107 ©Nigel Voaden; p.108 ©Pavel_Voitukovic/ shutterstock.com; p.110 (top), 121 ©Alexey Yakovlev/ Wikimedia Commons [CC BY-SA 2.0 (https://creativecommons. org/licenses/by-sa/2.0)]; p.110 (bottom) ©David Doussard; p.113(top) ©Miron Karlinsky; p.116(top,right) ©Doug Wechsler/ naturepl.com; p.117(top) © Steve and Alison Pearson Airlie Beach Qld Australia; p.118 ©James Wood; p.119(bottom) ©Rhett A. Butler/ Mongabay; p.123 © Michael Boppré; p.124 ©Paul Bertner/naturepl.com; p.125 ©Andy Sands/naturepl. com; p.128 (left) ©Egbert Friedrich; p.129 ©Paolo Mazzei; p.131, 167,195 ©Alberto Zilli; p.132 © David Agassiz ; p.133 © Egle Vičiuvienė; p.138, 165(bottom) ©Robert Thompson/ naturepl.com; p.139 ©Mark Heighes/shutterstock.com; p.140 © Ratnasingham, S. & Hebert, P. D. N. (2007). BOLD: The Barcode of Life Data System (www.barcodinglife.org). Molecular Ecology Notes 7, 355-364. DOI: 10.1111/j.1471-8286.2006.01678.x; p.141 © Image: Martin Heffer © Plant & Food Research; p.142(top) ©jdwfoto/shutterstock.com; p.142(bottom) ©Ted Lee Eubanks; p.143 ©Steve Ogden; p.144(left), 173(bottom, left) ©Shipher Wu;(right) ©Jose Amorin; p.145(top) ©Duncan Mcewan/naturepl.com; (bottom, left), 170(bottom) ©George Gibbs; p.145 (bottom, right) ©Bob Heckford; p.146 © Loetti, Veronica, Valverde, Alejandra, & Rubel, Diana Nora. (2016).

Galls of Cecidoses eremita Curtis and Eucecidoses minutanus Brèthes (Lepidoptera: Cecidosidae) in Magdalena, Buenos Aires Province: preliminary study and associated fauna. Biota Neotropica, 16(4), e20160161; p.148 © Pavel Gorbunov; p.149 © School of Biological Sciences, The University of Hong Kong; p.151(top) ©Dan Young; p.151(bottom) © ike Beauregard from Nunavut, Canada [CC BY 2.0 (https://creativecommons.org/ licenses/by/2.0)]; p.153 © Grehan, John. (2018). Ghost moth fragments of Gondwana; p.154 ©David Demergès; p.155 © Muséum National d'Histoire Naturelle; p.159 Classiccardinal [CC BY-SA 4.0 (https://creativecommons.org/licenses/by-sa/4.0)]; p.162(top) ©Tolson Museum; p.162(bottom) ©ajt/shutterstock. com; p.164(left) ©Lorraine Bennery/naturepl.com; (right) ©Serguei Koultchitskii/shutterstock.com; p.166 © Natalia Zakhartseva; p.173(top) © Phuket Nature Tours/Ian Dugdale and Punjapa Phretsri; p.173(bottom, right) ©Simon Shim/ shutterstock.com; p.175 © Zoological Journal of the Linnean Society/Shen-Horn Yen; p.176(top) Zhang W, Shih C, Labandeira CC, Sohn J-C, Davis DR, Santiago-Blay JA, et al. (2013) New Fossil Lepidoptera (Insecta: Amphiesmenoptera) from the Middle Jurassic Jiulongshan Formation of Northeastern China. PLoS ONE 8(11); p.178 © Johns, C et al Evidence of an Undescribed, Extinct Philodoria Species (Lepidoptera: Gracillariidae) from Hawaiian Hesperomannia Herbarium Specimens), 2014; p.179 ©Patrick Landmann/Science Photo Library; p.180 ©Ann Ronan Picture Library/Heritage Images/Science Photo Library; p.182 ©topimages/shutterstock.com; p.184(top) ©Sandra Moraes/shutterstock.com; p.184(bottom) Jean-Pol GRANDMONT [CC BY-SA 3.0 (https://creativecommons.org/ licenses/by-sa/3.0)]; p.185(bottom) ©Santiage David-Rivera; p.187(left) ©Mel Bray; p.187(right) ©Paul Harcourt Davies/ naturepl.com; p.189 © Heinrich-Barth-Institute, author Mary-Theres-Erz; p.190 (top)NH53/Flickr;p.190(bottom) ©Anan Kaewkhammul/shutterstock.com; p.191(right) ©Doo-Hyun Lo et al, 2011, Biomimetic microlens array with antireflective "moth-eye" surface, Soft Matter, 7: 6404–6407; p.191(right) César Hernández/CSIC; p.194 ©Christine Miller; p.195(top) ©Aaskolnick/Dreamstime.com

Unless otherwise stated images ©The Trustees of the Natural History Museum, London.
Every effort has been made to contact and accurately credit all copyright holders. If we have been unsuccessful, we apologise and welcome correction for future editions and reprints.